Transformative Natural Language Processing

Akshi Kumar • Saurabh Raj Sangwan
Editors

Transformative Natural Language Processing

Bridging Ambiguity in Healthcare, Legal, and Financial Applications

Springer

Editors
Akshi Kumar
School of Computing
Goldsmiths, University of London
London, UK

Saurabh Raj Sangwan
School of Computer Science and Engineering, Artificial Intelligence and Machine Learning
G L Bajaj Institute of Technology and Management
Greater Noida, Uttar Pradesh, India

ISBN 978-3-031-88987-5   ISBN 978-3-031-88988-2 (eBook)
https://doi.org/10.1007/978-3-031-88988-2

This Springer imprint is published by the registered company Springer Nature Switzerland AG
The registered company address is: Gewerbestrasse 11, 6330 Cham, Switzerland

Preface

Natural Language Processing (NLP) has emerged as a transformative tool for addressing the complexities of language interpretation, analysis, and generation. Its applications span critical sectors such as healthcare, law, and finance where ambiguity and precision play pivotal roles. As these high-stakes domains increasingly adopt and rely on NLP solutions, a deeper exploration of practical implementations, ethical considerations, and innovative advancements becomes essential.

This book offers a comprehensive look into how NLP is reshaping these sectors. Through ten chapters, it highlights cutting-edge applications, the challenges faced, and future opportunities.

- **Chapter 1: Introduction to Natural Language Processing in High-Stakes Domains** introduces fundamental NLP principles, key technologies like transformers, and critical challenges, including ambiguity, data privacy, and bias. It sets the foundation for understanding NLP's potential in complex fields.
- **Chapter 2: NLP in Medicine: Enhancing Diagnostics and Patient Care** examines how NLP analyzes unstructured medical data to enhance diagnostics and decision-making while addressing ethical issues such as patient privacy.
- **Chapter 3: NLP in the Legal Domain: Ensuring Precision and Compliance** explores its role in automating document analysis, predicting litigation outcomes, and enhancing compliance. It tackles the challenges posed by the complexity of legal language.
- **Chapter 4: Introduction to NLP in Finance: Sentiment Analysis and Risk Management** focuses on sentiment analysis, fraud detection, and market analysis. It showcases how NLP supports real-time decision-making and mitigates risks.
- **Chapter 5: Managing Uncertainty in NLP: Advanced Techniques and Approaches** addresses the inherent ambiguity of language, presenting advanced techniques like Bayesian methods and calibration approaches for building robust systems.
- **Chapter 6: NLP for Fraud Detection and Security in Financial Documents** highlights the integration of NLP and machine learning to detect financial

anomalies and fraud through innovative techniques like word embeddings and contextual language models.
- **Chapter 7: Multilingual and Cross-Linguistic Challenges in NLP** discusses strategies for building inclusive systems that address linguistic diversity and cater to underserved languages, fostering global equity.
- **Chapter 8: NLP in Action: Case Studies from Healthcare, Finance, and Industry** provides real-world case studies across industries like education, retail, and government, illustrating NLP's transformative impact.
- **Chapter 9: Generative Large Language Models in Clinical, Legal and Financial Domains** explores tools (like GPT), which enable advanced content creation and conversational AI. It examines their transformative potential in high-stakes domains while addressing ethical concerns.
- **Chapter 10: Responsible and Ethical AI in Natural Language Processing** emphasizes transparency, fairness, and inclusivity. It offers practical frameworks to ensure ethical development and deployment of NLP systems.

This book is a testament to the profound impact of NLP in overcoming language challenges across critical sectors. By combining cutting-edge research with practical applications, it charts a path forward for responsible and transformative innovation, inspiring readers to envision and contribute to a future where NLP technologies empower humanity with precision, inclusivity, and ethical integrity.

Contents

Chapter 1
Introduction to Natural Language Processing in High-Stakes Domains

Akshi Kumar and Saurabh Raj Sangwan

Abstract Natural Language Processing (NLP) has emerged as a critical field within artificial intelligence (AI), transforming human-computer interactions by enabling machines to understand, interpret, and generate human language. Its applications span high-stakes domains such as healthcare, finance, legal services, education, and national security, where accuracy, reliability, and ethical considerations are paramount. In these sectors, NLP enhances decision-making, operational efficiency, and service personalization by analyzing vast amounts of unstructured data, from clinical notes to legal contracts. This chapter examines the role of NLP in these high-stakes environments, highlighting key technologies such as transformers, named entity recognition (NER), and domain-specific transfer learning. It also addresses the unique challenges NLP faces in these contexts, including data privacy, bias, interpretability, and regulatory compliance. Through rigorous testing and development, NLP can provide ethical, transparent, and impactful solutions that meet the critical demands of these domains.

Natural Language Processing (NLP) is a branch of artificial intelligence (AI) that deals with the interaction between computers and human (natural) languages [1]. In simpler terms, it's about teaching machines to understand, interpret, and respond to human language in a way that's useful and meaningful. This could be in the form of written text or spoken language, allowing machines to process and generate responses as if they were human. Think of NLP as giving computers the ability to "*read*" and "*talk*" in the same way humans do, without needing structured, pre-defined inputs like traditional computer languages. Unlike traditional programming languages, which rely on structured and explicit instructions, human languages are

A. Kumar (✉)
School of Computing, Goldsmiths, University of London, London, UK
e-mail: Akshi.Kumar@gold.ac.uk

S. R. Sangwan
Department of Computer Science and Engineering, Artificial Intelligence and Machine Learning, G L Bajaj Institute of Technology and Management,
Greater Noida, Uttar Pradesh, India
e-mail: saurabh.sangwan@glbitm.ac.in

A. Kumar, S. R. Sangwan (eds.), *Transformative Natural Language Processing*,
https://doi.org/10.1007/978-3-031-88988-2_1

inherently ambiguous and complex, making NLP a challenging yet crucial domain in AI. NLP systems are designed to bridge the gap between human communication and computer understanding by processing, analyzing, and deriving meaning from natural language inputs. Imagine trying to explain a joke to a computer. Humans can understand a punchline because they pick up on cultural references, shared knowledge, or even tones like sarcasm or irony. However, computers don't have this implicit understanding; they need structured, logical rules. When a human says, "*I'm feeling blue*," they could mean they're sad, but the literal interpretation of the colour blue would confuse a machine without the proper context. NLP systems aim to help machines understand these subtleties in language.

The significance of NLP lies in its ability to enhance human-computer interaction by allowing machines to comprehend and communicate using human language. Traditional computer systems operate on structured inputs like code, but NLP enables the integration of unstructured data—such as text, speech, and even images containing linguistic information—into computational systems. This ability to process and interpret unstructured data is essential in a world where vast amounts of information are communicated through language, whether in emails, research articles, social media posts, or spoken interactions. NLP allows machines to analyze and interpret this data, providing valuable insights, automating processes, and enhancing the efficiency of numerous sectors, including healthcare, finance, education, and law. By giving machines the ability to understand human language, NLP opens opportunities for innovation, making AI systems more intuitive, accessible, and effective.

1.1 How NLP Mimics Human Language Understanding and Processing?

Human language is not just about stringing words together in the correct order. Understanding it is complex for both people and machines. When humans communicate, they rely on several layers of understanding, such as knowing the words, the rules of grammar, and the context of the conversation. Machines, on the other hand, need to rely on structured algorithms to perform these tasks.

At the most fundamental level, human language understanding starts with parsing syntax and applying grammatical rules. Just as humans learn these rules implicitly over time, NLP systems begin by breaking down sentences into their structural components using algorithms. This process is known as syntactic analysis or parsing, where NLP tools identify parts of speech (such as nouns, verbs, and adjectives) and establish relationships between these elements to understand sentence dependencies [2]. For example, in the sentence "*The cat sat on the mat*," an NLP system would recognize "*cat*" as the subject, "sat" as the verb, and "mat" as the object, analyzing how these parts fit together to form meaning. Beyond simple grammar, syntax parsing is vital for understanding more complex linguistic structures such as subordinate clauses, passive voice, or nested sentences. Humans intuitively handle such complexities, but machines must rely on formal syntactic rules. Advanced parsing algorithms, such as those using context-free grammars (CFG) or dependency parsing, help machines understand how different parts of a sentence relate to one another hierarchically [3].

While syntactic analysis focuses on structure, semantic understanding deals with meaning. Humans don't just understand words individually; they infer meaning based on context, shared knowledge, and the relationships between words. For instance, humans recognize that "bank" could refer to a financial institution or the side of a river, depending on the surrounding words in the sentence. NLP systems aim to mimic this through semantic analysis. Semantic understanding in NLP is often achieved through models like word embeddings, which transform words into vectors of numbers based on their contextual usage in large datasets [4]. Techniques such as Word2Vec [5] and GloVe (Global Vectors for Word Representation) [6] capture word meanings by analyzing how words co-occur with others in massive corpora of text. In these representations, words with similar meanings (e.g., "king" and "queen" or "happy" and "joyful") will have similar vectors, allowing the model to infer relationships between them. For more advanced semantic tasks, such as handling metaphors, humour, or implied meaning, deep learning models like transformers come into play. These models, including BERT (Bidirectional Encoder Representations from Transformers) [7] and GPT (Generative Pretrained Transformer) [8], use vast amounts of training data to develop a deep understanding of language. Transformers operate on the principle of attention mechanisms, which allow them to focus on different parts of a sentence and interpret words in relation to their context. This allows the model to understand nuances like sarcasm ("*Oh, great!*" meaning something is actually not great) or idiomatic expressions ("*kick the bucket*" meaning to die).

Another layer of complexity in human language is ambiguity [9]. Humans can easily resolve ambiguity based on the context in which a sentence is used. For example, the sentence "*She saw the man with the telescope*" can be interpreted in two ways: either "*she*" used the telescope to see the man, or the man had the telescope. Human listeners use contextual clues to resolve such ambiguities. NLP systems must use sophisticated techniques to do the same, often relying on probabilistic models that make educated guesses about which interpretation is most likely given the context. In situations where context is paramount, such as dialogue systems (e.g., virtual assistants like Alexa or Siri), contextual NLP models are employed to track and store the history of a conversation. For example, in a dialogue, when a user asks a follow-up question like "*What about tomorrow?*" after an initial question, NLP systems use context tracking to infer that "*tomorrow*" refers to a specific event discussed earlier, rather than a generic reference to the next day.

Human language is also filled with emotional tones, intent, and social cues. Humans can detect whether someone is angry, happy, or sarcastic based on subtle changes in wording, tone, or phrasing. NLP systems are now being trained to do the same using techniques like sentiment analysis [10, 11] and emotion detection [12]. For example, customer service chatbots use sentiment analysis to detect when a user is frustrated and may offer more empathetic responses or escalate the issue to a human agent. In healthcare, emotion detection models are used in mental health apps to analyze patient sentiment from text inputs and track emotional well-being over time.

In terms of intent, NLP systems must often detect what the user wants, even if it's not explicitly stated. For instance, when someone types, "*I need to book a flight*," an NLP system can infer that the user's intent is to find available flights, rather than asking for general information about airlines. This is known as intent classification, and it's crucial for applications like virtual assistants, chatbots, and recommendation systems.

1.2 Evolution of NLP Technologies

NLP has experienced significant evolution, transitioning from rule-based systems to advanced AI models that are now central to numerous applications. This journey is marked by four main phases: symbolic methods, statistical methods, deep learning, and the recent emergence of large language models (LLMs). Each phase has driven transformative advancements, enabling machines to process human language with increasing accuracy and depth. Today, NLP spans fields such as translation, sentiment analysis, conversational AI, and more. Figure 1.1 below illustrates this progression, highlighting milestones in each phase.

- **Symbolic Methods (1950s–1990s)**
- Early NLP systems relied on symbolic, rule-based approaches where linguistic rules were manually programmed. These methods used dictionaries, grammar rules, and fixed patterns to process language in a deterministic way. Early machine translation systems, for instance, depended on predefined grammatical rules to translate languages, which limited flexibility and scalability. Systems like ELIZA (1966) and SHRDLU (1970) demonstrated basic conversation and text manipulation, but they struggled with language's inherent ambiguity and variability [13]. As a result, symbolic methods couldn't handle complex, unstructured data effectively.

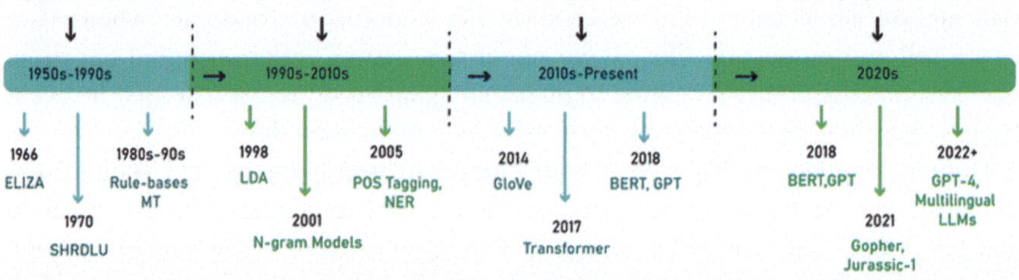

Fig. 1.1 The timeline of NLP technologies from symbolic methods to the rise of deep learning

- **Statistical NLP (1990s–2010s)**:
- The 1990s brought statistical models that allowed NLP systems to learn from large datasets, making them more flexible and scalable. This shift introduced probabilistic models such as Hidden Markov Models (HMMs) [14] and Conditional Random Fields (CRFs) [15], advancing capabilities in text and speech processing. Techniques like N-grams for next-word prediction, Part-of-Speech Tagging [16], and Named Entity Recognition (NER) [17] enabled systems to identify word categories and specific entities. Latent Dirichlet Allocation (LDA) allowed topic modelling [18], grouping related themes within large text collections. Statistical methods proved robust, especially for translation and speech recognition, but they still struggled with capturing deeper semantics and long-range dependencies in language.
- **Deep Learning (2010s–Present)**:
- The rise of deep learning in the 2010s marked a transformative phase in NLP. Neural networks, particularly Recurrent Neural Networks (RNNs) [19] and Long Short-Term Memory (LSTM) [20] networks, improved text pattern recognition without explicit rule-coding. Transformer-based architectures, including BERT (Bidirectional Encoder Representations from Transformers) and GPT (Generative Pretrained Transformer), enabled models to understand complex context and long-range dependencies, setting a new standard in language processing. Key innovations like word embeddings (e.g., Word2Vec, GloVe) facilitated semantic comprehension, representing words as multi-dimensional vectors. The transformer architecture introduced in the seminal "Attention is All You Need" paper [21] made modelling dependencies in language easier, leading to models like GPT-3 and BERT. Transfer learning also emerged, enabling these models to be fine-tuned with smaller, domain-specific datasets, enhancing applications in sentiment analysis, summarization, and creative writing.
- **Large Language Models (LLMs)**:
- Building on deep learning foundations, the development of large language models (LLMs) [22] like GPT-3, GPT-4, and Google's T5 marked a new era in NLP. These models, pre-trained on massive datasets, are capable of performing a wide range of tasks with minimal task-specific training, known as few-shot or zero-shot learning. With billions of parameters, LLMs possess unparalleled capacity for understanding, generating, and adapting to complex language structures, enabling near-human level performance in tasks such as summarization, question-answering, translation, and creative content generation. LLMs excel in capturing nuanced context, understanding idioms, and generating coherent, contextually appropriate responses, transforming fields like customer service, education, and healthcare [23].

Today, NLP technologies, especially LLMs, enable machines to tackle complex language tasks with unprecedented accuracy, establishing a solid foundation for future advancements in language-based AI applications.

1.3 Real-Time Examples of NLP in Action

NLP is used in a wide range of applications, from virtual assistants like Siri and Alexa to machine translation services such as Google Translate, and from text-based sentiment analysis to complex medical diagnostic tools. Its versatility has made it an integral component of modern AI systems. Moreover, it is integral to the development of intelligent agents and autonomous systems that can perform complex tasks, from diagnosing diseases to offering legal advice, with greater accuracy and efficiency than ever before. NLP is not just a theoretical concept—it's already embedded in many applications we use every day. Some of the most common examples include:

- **Virtual Assistants (Siri, Alexa, Google Assistant)**: Virtual assistants are one of the most visible applications of NLP [24]. These tools use NLP to understand spoken commands and respond appropriately. For instance, when you ask Siri to "*set a reminder for tomorrow at 9 AM*," NLP helps the assistant understand that "*set a reminder*" is the command, "*tomorrow*" refers to the next day, and "*9 AM*" is the time. The system processes the request and schedules the reminder. These assistants continue to evolve and become more conversational, thanks to advances in NLP.
- **Autocorrect in Smartphones**: When you type a sentence on your phone, the autocorrect system uses NLP to understand the context of your text [25]. If you type "*Their going to the store*," it suggests changing "*their*" to "*they're*," because it recognizes that "*they're*" fits better in the sentence's structure.
- **Machine Translation (Google Translate)**: Google Translate[1] uses NLP to break down text from one language and reassemble it in another. In the past, early translation models relied on word-for-word translations, which often resulted in awkward or incorrect translations. Today's translation models powered by NLP take context into account, providing more fluent and accurate translations. For example, translating idiomatic phrases like "*it's raining cats and dogs*" into another language can be tricky, but NLP models are trained to recognize these expressions and provide contextually appropriate translations.
- **Chatbots in Customer Service**: Many companies deploy chatbots on their websites to help with customer inquiries. These chatbots use NLP to understand user queries and provide relevant responses. If a customer types "*I need help with my order*," the chatbot can recognize that the user needs assistance with an order and respond with appropriate options like tracking an order, refund policies, etc. Chatbots like this are used by companies such as Amazon, Uber, and banks to streamline customer interactions.
- **Sentiment Analysis (Social Media Monitoring)**: NLP powers sentiment analysis tools that analyze large amounts of text to determine the emotional tone. This is widely used by companies to gauge customer sentiment on social media

[1] https://translate.google.com/?sl=auto&tl=en&op=translate

platforms like Twitter or Facebook. For example, when a company wants to know how people feel about a new product, sentiment analysis can quickly scan thousands of social media posts and categorize them as positive, negative, or neutral. This helps companies understand the public's reaction without manually reading through every post.

- **Spam Detection in Email**: Email services like Gmail use NLP algorithms to filter out spam emails. These algorithms analyze the text in incoming messages to determine whether they contain spam-related content like phishing attempts or unwanted advertisements. Over time, NLP models improve by learning from new examples of spam, making them increasingly accurate.
- **Healthcare Diagnostics**: NLP is being used in the healthcare industry to analyze patient records, medical research papers, and clinical notes to help doctors make better diagnoses. For instance, IBM's Watson for Oncology [26] uses NLP to read through vast amounts of medical literature and help doctors recommend treatment plans for cancer patients. NLP helps extract relevant information from unstructured medical data (like doctors' notes) and transforms it into actionable insights.

NLP's broad range of applications is one of the key reasons it has become integral to modern AI systems, with significant impact across various fields. In the financial sector, NLP is widely used for tasks such as analyzing financial news, extracting insights from reports, and detecting fraud [27]. Banks leverage NLP systems to identify potentially fraudulent transactions by analyzing patterns in customer communications [28]. In education, NLP powers intelligent tutoring systems that can understand student queries, assess responses, and provide personalized feedback [29, 30]. It also plays a crucial role in grading essays by evaluating the structure and content of written submissions. In the legal field, NLP helps streamline the review and interpretation of large volumes of legal documents [31]. Tools like ROSS Intelligence [32] allow lawyers to quickly find relevant case law and legal precedents, making legal research more efficient. Governments and defense agencies also rely on NLP to analyze intelligence reports and detect threats in real-time. For example, NLP is employed to monitor social media for potential security risks or track communications for criminal activity, demonstrating its critical role in safeguarding national security.

1.4 High-Stakes Domains: Definition and Characteristics

High-stakes domains are fields where the accuracy, reliability, and ethical considerations of technology play a critical role in decision-making and outcomes. In these sectors, mistakes or oversights can lead to significant consequences, including financial losses, legal repercussions, harm to public health, or even loss of life. Typical examples of high-stakes domains include healthcare, finance, legal services, education, and national security. In each of these areas, the stakes for precision, ethical compliance, and transparency are extremely high, as errors could negatively impact individuals or society on a large scale. High-stakes domains share several defining characteristics:

- **Critical Impact**: Decisions in these fields have profound consequences. For example, in healthcare, an incorrect diagnosis or inappropriate treatment plan suggested by an AI model can harm patients. In finance, a minor misinterpretation of market trends or risk could result in massive financial losses.
- **Need for Precision and Accuracy**: Accuracy is paramount in high-stakes domains. Systems must interpret information and derive insights with a high degree of precision. For instance, legal NLP applications need to accurately understand and summarize legal documents, as a minor misinterpretation could affect case outcomes.
- **Ethical and Regulatory Considerations**: High-stakes sectors are often subject to strict regulations and ethical guidelines. Healthcare, finance, and legal fields require adherence to laws protecting personal information and ensuring unbiased, fair treatment. NLP systems used in these fields must comply with these regulations and incorporate mechanisms to ensure ethical practices, such as handling sensitive data securely and avoiding bias in decision-making.
- **Trust and Transparency**: Trust is essential in high-stakes domains. Users and stakeholders must be able to rely on AI systems, knowing that these technologies will perform consistently and accurately. Transparency in NLP models is also crucial, as stakeholders need to understand how decisions are made, especially in cases where outcomes are legally or ethically binding.
- **Requirement for Explainability**: In high-stakes applications, it's often necessary to explain how a model arrived at a particular decision or recommendation. For instance, in law enforcement, NLP systems used to analyze crime patterns must provide transparent reasoning for their findings. Explainability becomes even more critical in cases where stakeholders need to make informed decisions based on AI suggestions.

These characteristics make high-stakes domains challenging environments for NLP systems, requiring advanced methodologies and rigorous testing to ensure reliability, transparency, and compliance.

1.5 The Role of NLP in High-Stakes Domains

In high-stakes domains, Natural Language Processing (NLP) enables the automated processing, analysis, and comprehension of vast amounts of unstructured language data, which is crucial for informed decision-making and operational efficiency. Sectors such as healthcare, finance, legal, education, and national security often rely on extensive documentation, text records, and communication channels. NLP technologies facilitate fast and accurate insights, supporting human oversight in areas where time and precision are critical. By enabling machines to interpret and respond to human language, NLP enhances workflows, making operations faster, more consistent, and less prone to error.

- **Healthcare**: In healthcare, NLP plays a vital role in analyzing medical records, clinical notes, and research articles, enabling more efficient diagnostic and treatment processes. This unstructured data contains essential patient information but interpreting it manually can be overwhelming. NLP systems, such as IBM Watson Health, scan thousands of medical documents quickly, assisting doctors in identifying treatment options for complex cases. For instance, NLP-driven diagnostic tools extract critical details from patient records, helping clinicians see patterns indicative of health risks or underlying conditions. This can lead to more accurate preventative care, improved treatment efficacy, and well-informed clinical decisions based on a complete understanding of patient history. Additionally, NLP can summarize the latest medical research, helping doctors keep up with the rapidly evolving medical literature and offering insights that might otherwise go unnoticed. NLP also enhances the patient experience by analyzing feedback and satisfaction surveys, extracting sentiments and identifying trends that hospitals can use to improve services. Furthermore, predictive analytics driven by NLP models can anticipate patient outcomes, such as readmission risks or recovery times, allowing healthcare providers to deliver proactive, personalized care.
- **Finance**: The financial sector relies heavily on real-time information and accurate trend prediction, and NLP provides significant advantages by analyzing various data sources, including financial reports, news articles, and social media sentiment, to predict market behaviour. Sentiment analysis, for instance, can help identify public opinion on companies or financial instruments, guiding investors and traders in decision-making. NLP-powered risk assessment tools analyze communication data to detect patterns or anomalies, thus safeguarding financial systems from potential fraud. Fraud detection algorithms use NLP to identify suspicious language or actions in financial transactions, adding a layer of security and protecting clients. In customer service, NLP-driven chatbots handle common banking inquiries, improving response times and customer satisfaction. These chatbots use NLP to analyze conversation patterns and ensure compliance with financial standards, further automating customer interactions. NLP also plays a crucial role in regulatory compliance by processing legal and regulatory texts, helping financial institutions stay updated on requirements and ensure adherence, thereby reducing risk.
- **Legal**: In legal contexts, NLP assists in managing and analyzing large volumes of legal documents, contracts, and case files. Reviewing legal documentation manually is time-intensive, and NLP-powered tools, like ROSS Intelligence, streamline this process by quickly summarizing case law, contracts, and legal statutes. These tools allow lawyers to locate relevant information more efficiently, freeing them from repetitive tasks. Additionally, NLP algorithms can scan legal language for inconsistencies or clauses that require closer scrutiny, reducing the risk of oversight and increasing accuracy in legal research. NLP-

powered e-discovery tools aid in the evidence-gathering process by automatically scanning through vast datasets for pertinent information, making case preparation more thorough and less labour-intensive. In contract lifecycle management (CLM), NLP identifies risky clauses, tracks obligations, and ensures legal compliance across contract amendments, making it invaluable for law firms and legal departments. By automating these repetitive but essential tasks, NLP allows legal professionals to focus on higher-level strategy and decision-making, ultimately leading to better client service and compliance.

- **Education**: In the field of education, NLP-powered intelligent tutoring systems enhance teaching and learning experiences through personalized interactions and real-time feedback. These systems respond to student queries, grade assignments, and provide tailored feedback based on individual performance. For instance, essay-grading systems analyze grammar, structure, and sentiment, helping educators evaluate student submissions more efficiently and objectively. Beyond grading, sentiment analysis gauges student satisfaction and engagement, enabling educational institutions to tailor their approach and improve learning outcomes. NLP-driven automatic question generation tools create quizzes or assignments from course material, which aids instructors and provides students with an interactive way to review content. NLP also plays a role in maintaining academic integrity through plagiarism detection systems, which analyze student submissions to identify similarities and ensure originality.
- **National Security**: In national security, NLP systems monitor and analyze data from sources such as social media, news, and intercepted communications, supporting real-time threat detection. By analyzing vast amounts of text data, NLP can detect linguistic cues that indicate potential criminal or terrorist activities. For instance, NLP-based systems identify specific patterns or keywords that suggest rising tensions or malicious intent. This capability is critical for government agencies and law enforcement in preempting and responding to threats. In cybersecurity, NLP helps detect early signs of cyber threats by analyzing threat reports, open-source intelligence, and communication patterns. Furthermore, during emergencies or natural disasters, NLP enables real-time analysis of social media and public data, providing valuable information for crisis management and coordination. NLP's language recognition and translation features also enable cross-language threat analysis, allowing security agencies to monitor communication in multiple languages.

In each high-stakes domain, NLP applications are becoming increasingly sophisticated, enabling organizations to streamline processes, improve decision-making, and deliver tailored solutions. From enhancing patient care in healthcare to ensuring compliance in finance, NLP tools are transforming how critical information is processed and acted upon. Below are specific use cases that highlight the depth and versatility of NLP's role across key sectors:

- Healthcare:

(a) *Patient Experience and Feedback Analysis:* NLP can process patient feedback from online reviews or satisfaction surveys, extracting sentiments and identifying areas where healthcare services need improvement. Such insights can guide hospitals in tailoring patient care and addressing specific concerns.
(b) *Predictive Analytics for Patient Outcomes:* By analyzing clinical notes and electronic health records, NLP can help predict patient outcomes, such as readmission risks, length of hospital stays, or post-surgery recovery time, allowing proactive care.

- Finance:

(a) *Automated Customer Support in Banking:* NLP chatbots handle common inquiries (e.g., balance checks, transaction details) for banks, improving customer service speed and satisfaction. NLP models also help analyze chat transcripts to ensure compliance with financial standards.
(b) *Risk Assessment and Compliance Monitoring:* NLP-powered tools can analyze regulatory texts and financial documents to ensure firms stay compliant with industry regulations, automating what were once manual audits.

- Legal:

(a) *E-discovery and Evidence Analysis:* NLP systems assist in e-discovery by scanning vast amounts of legal documents for relevant evidence, which can be essential in case preparation and reduce manual workload.
(b) *Contract Lifecycle Management (CLM):* NLP streamlines contract reviews, renewals, and amendments, flagging risky clauses, tracking obligations, and ensuring legal compliance across contract lifecycles.

- Education:

(a) *Automatic Question Generation for Assessments*: NLP tools generate quiz questions or assignments from course material, assisting teachers and allowing students to review content in an engaging way.
(b) *Plagiarism Detection and Content Moderation*: NLP helps universities detect plagiarism in student submissions and ensures academic integrity by identifying unoriginal content or flagged phrases.

- National Security:

(a) *Cross-Language Threat Analysis:* NLP allows security agencies to analyze foreign language content, such as intercepted communications, by translating and analyzing for threats across languages.
(b) *Disaster Response Coordination:* During natural disasters or large-scale emergencies, NLP helps analyze public and social media data to identify crisis areas, coordinating timely responses.

1.6 Key Technologies and Models in High-Stakes NLP

In high-stakes domains, sophisticated NLP models and techniques are essential to ensure that systems perform reliably, accurately, and consistently under complex, variable conditions. These technologies not only improve contextual understanding and precision but also provide the flexibility needed to adapt to specific requirements in fields like healthcare, finance, legal, education, and security. The following are some of the most impactful models and techniques currently used in high-stakes NLP applications:

Transformer Models (e.g., BERT, GPT): Transformers have revolutionized NLP by introducing deep learning models that excel at understanding context and capturing intricate language structures. Two notable transformer models, BERT (Bidirectional Encoder Representations from Transformers) and GPT (Generative Pre-trained Transformer), have transformed how language is processed and understood.

- **BERT**: BERT is a bidirectional model, meaning it reads text in both directions, which enables it to capture subtle nuances and complex relationships between words. This bidirectional nature is especially valuable for high-stakes tasks, such as sentiment analysis in healthcare notes or question-answering in medical literature, where understanding the full context is critical. For instance, BERT is used to interpret clinical terms accurately by reading text from left to right and right to left, allowing it to capture complex medical or financial terminology.
- **GPT**: GPT is a generative model designed for producing coherent and contextually relevant text. In high-stakes domains, GPT is often used for summarization, data generation, and even drafting responses in customer service scenarios. Its ability to create human-like text is invaluable in legal settings, where summarizing case studies or generating initial document drafts can save time and ensure thoroughness.
- **Named Entity Recognition (NER)**: NER is a foundational NLP technique that focuses on identifying and classifying entities within text, such as names, locations, dates, and specific terms. In high-stakes domains, NER is essential for extracting critical information with precision. In healthcare, NER identifies specific drugs, symptoms, and treatments from clinical notes or medical records, making it easier for healthcare providers to extract vital patient information quickly. For example, an NER system can highlight drug names and dosages, helping doctors avoid prescription errors and ensuring patients receive accurate care. In legal contexts, NER plays a crucial role in document review by identifying key legal terms, case names, and clauses. This is essential for efficient legal research, as it allows attorneys to locate pertinent information quickly within extensive documentation. Furthermore, NER supports compliance by flagging terms or entities that might require legal verification, such as clauses in contracts or specific legal precedents. In finance, NER helps identify companies, locations, financial instruments, and other relevant entities within market reports or

economic news. This is particularly useful for risk assessment and tracking significant events that impact financial markets.

- **Transfer Learning for Domain-Specific NLP**: Transfer learning has become a critical tool in high-stakes NLP by enabling pre-trained NLP models to be fine-tuned for specific domains. By leveraging large, general datasets to pre-train models and then fine-tuning them on domain-specific data, transfer learning makes it possible for NLP systems to adapt to the unique language and terminology of high-stakes fields. In healthcare, transfer learning allows models to interpret complex medical terminology accurately, which is critical for tasks like diagnostic prediction, patient data extraction, and treatment recommendation. For example, a model pre-trained on general language data can be fine-tuned on clinical notes to better understand terms like "myocardial infarction" instead of just "heart attack." In finance, transfer learning enables NLP systems to process financial jargon and understand nuances in economic discussions. For instance, a model fine-tuned on financial data can better interpret terms like "asset-backed securities" or "quantitative easing," allowing it to provide more accurate analysis for market predictions or financial sentiment. In legal applications, transfer learning helps models understand legal terminology, which often includes highly specialized and archaic language. Fine-tuning on legal texts enables NLP systems to accurately interpret legal documents, draft contracts, or assist in case law analysis with a greater degree of relevance and reliability.
- These technologies enable NLP systems to deliver relevant, accurate, and insightful information across high-stakes applications. By building on transformers, leveraging entity recognition, and employing transfer learning for domain-specific adaptation, NLP models can address the unique challenges of each domain, making them essential tools for modern high-stakes NLP applications.

The Table 1.1 below provides a summary of NLP's role across these high-stakes domains, detailing its applications and benefits.

In each domain, NLP enhances operational efficiency, provides timely insights, and supports data-driven decisions. By streamlining tasks that would otherwise require human intervention, NLP allows professionals to focus on complex problem-solving, reducing errors and boosting organizational effectiveness.

1.7 Key Challenges in High-Stakes NLP Applications

NLP applications in high-stakes domains face unique challenges due to the critical nature of these fields. The need for accuracy, transparency, and ethical considerations is heightened in these contexts, where mistakes can lead to severe consequences. Below are some of the primary challenges and examples illustrating the impact of these issues.

Table 1.1 NLP applications and benefits in high-stakes domains

Domain	Role of NLP	Examples
Healthcare	Analyzes medical records, clinical notes, and research for insights; aids in diagnostics	IBM Watson Health recommends treatments, diagnostic tools extract patient history for risk prediction
Finance	Performs market analysis, sentiment detection, and fraud detection; informs investment decisions	Sentiment analysis of news/social media for market trends, fraud detection in transaction patterns
Legal	Reviews and summarizes legal documents, case files, and contracts; automates legal research	ROSS Intelligence assists in case law search, automated contract analysis saves time
Education	Powers intelligent tutoring, responds to queries, grades assignments, and assesses satisfaction	Automated grading of essays, sentiment analysis gauges student satisfaction for improvements
National Security	Monitors social media and communication data for real-time threat detection and analysis	Identifies language patterns indicating threats, flags security risks from social media monitoring

- **Data Privacy and Security**: NLP systems in high-stakes sectors often process highly sensitive information, such as medical records, financial data, or personal identifiers. Ensuring data privacy is crucial, as mishandling such information can lead to legal repercussions and a loss of public trust. Regulations like the General Data Protection Regulation (GDPR) in the EU and the Health Insurance Portability and Accountability Act (HIPAA) in the U.S. mandate strict standards for data security and privacy. For instance, in healthcare, NLP models need to be designed with patient confidentiality in mind, requiring robust data anonymization and encryption techniques. Models that process patient records for diagnostic support must strip personally identifiable information (PII) and ensure that sensitive data cannot be re-identified, even if intercepted.
- **Bias and Fairness**: High-stakes NLP applications need to operate fairly and without bias, as biased outcomes can lead to ethical violations and discrimination. Bias in NLP models often arises from biased training data, which may reflect societal stereotypes or imbalances. For example, in the legal domain, if an NLP model used for predicting case outcomes is trained on data that disproportionately reflects judgments against certain demographic groups, it may perpetuate or even exacerbate these biases. Similarly, in healthcare, a biased model could lead to incorrect diagnoses for underrepresented groups, resulting in unequal treatment. Addressing bias involves careful selection of training data, frequent auditing, and employing debiasing techniques, such as fairness constraints, that help reduce discriminatory behavior in model predictions. For example, some healthcare NLP models are now being retrained with more diverse patient data to improve diagnostic accuracy across different ethnic groups.
- **Interpretability and Explainability**: Many high-stakes sectors require explainable AI, meaning that stakeholders must understand how a model arrives at its conclusions. This is particularly important in fields like healthcare, finance, and law, where decisions based on NLP predictions are legally binding or ethically

significant. For instance, if a financial analyst uses an NLP model to evaluate the risk associated with a loan, understanding the model's reasoning is essential to justify its decisions to regulatory bodies. However, NLP models, especially deep learning models, are often "black boxes," making it challenging to interpret how they reach conclusions. Developing interpretable NLP models or implementing explainability tools such as LIME (Local Interpretable Model-agnostic Explanations) or SHAP (SHapley Additive exPlanations), is crucial to build trust and ensure that decision-makers can validate and understand model predictions. In healthcare, an interpretable model could clarify why it recommended a particular treatment by highlighting key patient data points it used to arrive at its suggestion.

- **Handling Ambiguity and Context**: Human language is inherently ambiguous, and context is often necessary to interpret meaning accurately. In high-stakes applications, this challenge is amplified, as the interpretation may vary significantly based on specific cases or scenarios. For example, in legal applications, the interpretation of terms in a contract may differ depending on prior case law or the context in which the contract was written. An NLP system processing legal contracts must distinguish between varied interpretations to avoid misinterpretation. In healthcare, ambiguity may arise when interpreting symptoms described in unstructured clinical notes, where terms could carry different implications based on the patient's history. NLP systems must be capable of interpreting nuances, resolving ambiguities, and adapting to specific contexts to provide accurate and relevant information in these domains.
- **Scalability and Real-Time Processing**: High-stakes domains often require real-time insights, particularly in sectors like finance and national security, where timely decisions are critical. NLP systems in these domains must handle large volumes of text data, such as news feeds, market data, or social media posts, and provide real-time responses without sacrificing accuracy. For example, in finance, real-time market analysis requires NLP models to process global news and social media sentiment instantly to identify potential market trends. In national security, NLP models must analyze social media and intercepted communications rapidly to detect potential security threats or unfolding crises. This scalability challenge demands efficient algorithms and infrastructure, as high-stakes NLP applications cannot afford delays or high computational costs when lives or finances are at risk.
- **Ethical and Regulatory Compliance**: High-stakes NLP applications must comply with stringent industry-specific regulations and ethical standards, which often restrict how data is collected, processed, and used. For instance, in healthcare, NLP models must adhere to HIPAA regulations to protect patient data privacy. In finance, NLP systems must comply with Anti-Money Laundering (AML) and Know Your Customer (KYC) requirements, which dictate how customer information can be used and stored. Failure to comply can lead to significant financial and legal consequences. In legal contexts, NLP systems used to process sensitive documents must follow strict guidelines to avoid breaching confidentiality agreements or mishandling classified information. Meeting these

ethical and regulatory standards requires continual auditing and updating of models to align with evolving regulations.

- **Model Generalization and Adaptability**: High-stakes NLP applications need to generalize well across different cases, environments, and jurisdictions. For example, an NLP model trained on U.S. case law may not perform accurately in countries with different legal frameworks, as legal terminology, principles, and case precedents vary widely. Similarly, in healthcare, a model trained on Western medical data may not generalize well to other regions with differing health practices or patient demographics. In national security, NLP models must adapt to diverse languages and cultural contexts, especially when analyzing foreign communications for threat detection. Ensuring that NLP models are adaptable and accurate across varied settings requires domain-specific fine-tuning and testing, as well as incorporating diverse datasets that reflect the specific environments in which the models will be deployed.

These challenges highlight the importance of rigorous development, testing, and validation for NLP models in high-stakes domains. Researchers and practitioners must continuously address these obstacles to build systems that are reliable, fair, and transparent. By doing so, NLP can achieve its potential to deliver ethical, accurate, and impactful solutions across high-stakes applications.

1.8 Real-World Case Studies

Adding real-world case studies enhances understanding of how NLP is actively transforming high-stakes sectors. These examples illustrate NLP's impact on efficiency, decision-making, and accuracy in critical environments.

1.8.1 IBM Watson Health: Enhancing Medical Recommendations

IBM Watson Health is a pioneering example of how NLP transforms healthcare. Leveraging its advanced AI capabilities, Watson Health analyzes vast amounts of unstructured medical data, including clinical notes, research articles, and patient histories, to provide doctors with evidence-based treatment recommendations. By using NLP, Watson can quickly scan through thousands of medical studies and clinical trials, which would be nearly impossible for a human to accomplish in a reasonable timeframe. The system extracts relevant information, compares it with patient data, and suggests potential treatments, allowing healthcare providers to make informed decisions swiftly. This capability is especially valuable in oncology, where treatment recommendations can be highly individualized and where Watson Health has supported oncologists in formulating personalized cancer treatment plans based on the latest research.

- **Example Impact**: At the Manipal Comprehensive Cancer Center in Bangalore, IBM Watson Health has been used to assist oncologists in tailoring treatment for cancer patients, helping to identify suitable therapies by analyzing genomic data and patient medical records. By processing unstructured text data and clinical research, Watson Health has improved the quality of care and accelerated the time-to-decision process in cancer treatment.

1.8.2 ROSS Intelligence: Transforming Legal Research

ROSS Intelligence uses NLP to streamline legal research, offering a transformative tool for legal professionals. Built on IBM's Watson, ROSS is an AI-powered legal research assistant designed to process and analyze legal documents, case law, and statutes. By employing NLP, ROSS can sift through extensive legal databases to find relevant information, saving lawyers hours of manual research. The platform can respond to natural language questions, such as "What are the legal precedents for breach of contract in New York?" and retrieve cases, statutes, and legal precedents that answer the query accurately. This capability enables lawyers to conduct more thorough research in less time and increases their ability to build stronger cases based on comprehensive legal findings.

- **Example Impact**: The law firm BakerHostetler adopted ROSS Intelligence to support its bankruptcy practice, utilizing the AI tool to conduct legal research efficiently. By automating the search for legal precedents and reducing the time required to find pertinent case law, ROSS Intelligence has enhanced the firm's research accuracy and allowed its attorneys to focus on strategy and client service rather than time-consuming document review.

1.8.3 JP Morgan's COiN: Contract Intelligence for Financial Compliance

In finance, JP Morgan developed COiN (Contract Intelligence), an NLP-powered platform to review and analyze complex legal and financial documents. COiN automates the extraction of critical data points from contracts, a task that would traditionally require extensive human resources. The system leverages NLP to identify essential clauses, such as payment terms and liabilities, and flags any terms that may be non-compliant with regulatory standards. By automating this process, COiN has significantly reduced the time and cost associated with contract review while ensuring adherence to financial regulations.

- **Example Impact**: COiN has enabled JP Morgan to review over 12,000 commercial credit agreements annually, a task that would typically require 360,000 h of manual labour. By ensuring that contractual terms meet compliance standards, COiN has strengthened JP Morgan's risk management strategy and improved the efficiency of its compliance efforts.

1.9 Ethical Frameworks and Compliance Standards in High-Stakes Domains

Given the sensitive nature of high-stakes domains, ethical frameworks and compliance standards are critical to ensuring that NLP applications uphold privacy, security, and fairness. Below are key frameworks and guidelines specific to each field.

1.9.1 Healthcare: HIPAA Compliance

In healthcare, the Health Insurance Portability and Accountability Act (HIPAA) governs the handling of protected health information (PHI) in the U.S. NLP applications that process medical records, or patient data must ensure HIPAA compliance to protect patient confidentiality and prevent unauthorized access to PHI. This involves implementing data anonymization, encryption, and secure storage practices. For instance, healthcare NLP tools, like those used by IBM Watson Health, must de-identify patient information in medical records to maintain privacy standards and comply with HIPAA requirements. Moreover, healthcare NLP applications must allow access only to authorized personnel, ensuring that PHI remains secure throughout the data processing workflow.

1.9.2 Finance: GDPR and AML/KYC Regulations

In finance, compliance with data protection laws like the General Data Protection Regulation (GDPR) and Anti-Money Laundering (AML) requirements is paramount. GDPR mandates that organizations obtain explicit consent for data use, ensure data portability, and provide users with the right to be forgotten. NLP systems in finance must incorporate mechanisms to allow for these requirements, especially when processing customer data in tasks such as sentiment analysis or transaction monitoring.

Financial institutions are required to implement Know Your Customer (KYC) policies to prevent fraud and money laundering. NLP applications in this sector must analyze communication patterns and customer data to identify suspicious behaviour in compliance with AML regulations. JP Morgan's COiN, for instance, helps automate contract compliance by ensuring that legal terms meet financial regulations, thereby enhancing transparency and preventing potential regulatory violations.

1.9.3 Legal Sector: Compliance with Confidentiality and Fairness

The legal domain is bound by stringent confidentiality agreements and ethical guidelines that demand the utmost security in handling sensitive legal information. NLP systems used for legal research, such as ROSS Intelligence, must be developed with strong security protocols to protect client confidentiality and prevent unauthorized access. Additionally, NLP applications in law must address bias to ensure fairness, especially if used in legal decision-making or predictive analytics. Ensuring compliance in this domain also involves auditing NLP models to avoid perpetuating biases present in training data, as biased outcomes could unfairly impact individuals or groups within the legal system.

1.9.4 Education: FERPA and Student Privacy

In the educational sector, the Family Educational Rights and Privacy Act (FERPA) protects the privacy of student education records. NLP systems used for educational purposes, such as intelligent tutoring systems or grading tools, must comply with FERPA by securing student data and preventing unauthorized access. For instance, an NLP-powered essay-grading system must anonymize student identities and restrict access to authorized educators only. Compliance with FERPA ensures that students' educational records remain confidential and that NLP tools used in education do not compromise student privacy.

1.9.5 National Security: Ethical Use of Surveillance Data

In national security, ethical guidelines dictate that NLP applications used for surveillance or threat detection must adhere to strict guidelines to avoid misuse. For instance, NLP systems that monitor social media or communication data for security purposes must be transparent about the data collected and used. This requires ethical oversight to prevent excessive surveillance or violation of individual privacy rights. Moreover, these systems must be designed to minimize biases that could lead to disproportionate scrutiny of certain demographic groups, ensuring that national security NLP tools are both effective and just.

1.9.6 Integrating Ethical Frameworks into NLP Applications

Each of these domains emphasizes specific compliance standards that NLP applications must address to operate ethically. Implementing these frameworks involves:

- **Data Anonymization and Encryption**: Ensuring that sensitive information, such as patient records or financial data, is anonymized to protect user privacy.
- **Bias Audits and Fairness Constraints**: Regularly auditing NLP models for biases and applying fairness constraints to mitigate discrimination risks.
- **Access Control and Authorization**: Limiting access to NLP systems to authorized personnel, especially in healthcare, finance, and legal sectors.
- **Transparency and Explainability**: Enabling explainable AI in high-stakes domains so that decision-makers understand the basis for each model's recommendations.

By integrating these ethical frameworks and compliance standards, NLP applications can better serve high-stakes domains, meeting both operational and regulatory demands with integrity and accountability.

1.10 Conclusion

NLP has evolved to play a transformative role in high-stakes domains, offering tools that automate, analyze, and enhance critical decision-making processes. Across sectors like healthcare, finance, legal, education, and national security, NLP enables unprecedented levels of insight by processing unstructured data with accuracy and speed. Real-world implementations, such as IBM Watson Health's clinical recommendations, ROSS Intelligence's legal research, and JP Morgan's COiN for financial compliance, highlight NLP's capacity to streamline complex tasks, increase productivity, and improve outcomes. However, with these advancements come heightened ethical and regulatory responsibilities, particularly around data privacy, bias, transparency, and compliance. Ethical frameworks, including HIPAA, GDPR, AML, FERPA, and confidentiality standards, are essential for guiding the responsible use of NLP, ensuring that applications uphold privacy, fairness, and accountability.

Key technologies, such as transformers, NER, and transfer learning, have proven indispensable in adapting NLP for these specialized fields, providing systems with the ability to understand context, capture domain-specific language, and offer interpretable insights. Moving forward, the development of ethical, adaptable, and transparent NLP models will remain essential for sustaining trust and effectiveness in high-stakes applications. This chapter establishes a foundation for understanding both the transformative potential and the critical considerations of NLP in high-stakes domains, emphasizing the need for ongoing research and innovation to meet the complex demands of these sectors.

References

1. Fanni, S. C., Febi, M., Aghakhanyan, G., & Neri, E. (2023). Natural language processing. In *Introduction to Artificial Intelligence* (pp. 87-99). Cham: Springer International Publishing.
2. Srinivasa-Desikan, B. (2018). *Natural Language Processing and Computational Linguistics: A practical guide to text analysis with Python, Gensim, spaCy, and Keras*. Packt Publishing Ltd.
3. Sowmith, J., Dasari, L. A., Mamidi, N. K., & Belwal, M. (2024, June). Parsing Ambiguous Grammar-A Survey. In *2024 15th International Conference on Computing Communication and Networking Technologies (ICCCNT)* (pp. 1-7). IEEE.
4. Worth, P. J. (2023). Word embeddings and semantic spaces in natural language processing. *International journal of intelligence science*, *13*(1), 1-21.
5. Church, K. W. (2017). Word2Vec. *Natural Language Engineering*, *23*(1), 155-162.
6. Pennington, J., Socher, R., & Manning, C. D. (2014, October). Glove: Global vectors for word representation. In *Proceedings of the 2014 conference on empirical methods in natural language processing (EMNLP)* (pp. 1532-1543).
7. Devlin, J., Chang, M. W., Lee, K., & Toutanova, K. (2018). Bert: Bidirectional encoder representations from transformers. *arXiv preprint arXiv:1810.04805*, 15.
8. Achiam, J., Adler, S., Agarwal, S., Ahmad, L., Akkaya, I., Aleman, F. L., ... & McGrew, B. (2023). Gpt-4 technical report. *arXiv preprint arXiv:2303.08774*.
9. Yadav, A., Patel, A., & Shah, M. (2021). A comprehensive review on resolving ambiguities in natural language processing. *AI Open*, *2*, 85-92.
10. Kumar, A., & Teeja, M. S. (2012). Sentiment analysis: A perspective on its past, present and future. *International Journal of Intelligent Systems and Applications*, *4*(10), 1.
11. Kumar, A., & Albuquerque, V. H. C. (2021). Sentiment analysis using XLM-R transformer and zero-shot transfer learning on resource-poor indian language. *Transactions on Asian and Low-Resource Language Information Processing*, *20*(5), 1-13.
12. Kumar, A., Dogra, P., & Dabas, V. (2015, August). Emotion analysis of Twitter using opinion mining. In *2015 Eighth International Conference on Contemporary Computing (IC3)* (pp. 285-290). IEEE.
13. Kumar, E. (2013). *Natural language processing*. IK International Pvt Ltd.
14. Eddy, S. R. (1996). Hidden markov models. *Current opinion in structural biology*, *6*(3), 361-365.
15. Cohn, T. A. (2007). *Scaling conditional random fields for natural language processing*. University of Melbourne, Department of Computer Science and Software Engineering, Faculty of Engineering.
16. Martinez, A. R. (2012). Part-of-speech tagging. *Wiley Interdisciplinary Reviews: Computational Statistics*, *4*(1), 107-113.
17. Song, S., Zhang, N., & Huang, H. (2019). Named entity recognition based on conditional random fields. *Cluster Computing*, *22*, 5195-5206.
18. Jelodar, H., Wang, Y., Yuan, C., Feng, X., Jiang, X., Li, Y., & Zhao, L. (2019). Latent Dirichlet allocation (LDA) and topic modeling: models, applications, a survey. *Multimedia tools and applications*, *78*, 15169-15211.
19. Tarwani, K. M., & Edem, S. (2017). Survey on recurrent neural network in natural language processing. *Int. J. Eng. Trends Technol*, *48*(6), 301-304.

20. Van Houdt, G., Mosquera, C., & Nápoles, G. (2020). A review on the long short-term memory model. *Artificial Intelligence Review*, *53*(8), 5929-5955.
21. Vaswani, A. (2017). Attention is all you need. *Advances in Neural Information Processing Systems*.
22. Chang, Y., Wang, X., Wang, J., Wu, Y., Yang, L., Zhu, K., ... & Xie, X. (2024). A survey on evaluation of large language models. *ACM Transactions on Intelligent Systems and Technology*, *15*(3), 1-45.
23. Johnsen, M. (2024). *Large Language Models (LLMs)*. Maria Johnsen.
24. OLUSEGUN, J. TOPIC: Case Study: Enhancing User Interaction: Leveraging NLP for Contextual Understanding in Virtual Assistants.
25. "Ray" Zhang, M., Wen, H., Cui, W., Zhu, S., Andrew Schwartz, H., Bi, X., & Wobbrock, J. O. (2021). AI-Driven Intelligent Text Correction Techniques for Mobile Text Entry. *Artificial Intelligence for Human Computer Interaction: A Modern Approach*, 131-168.
26. Choi, Y. S. (2017). Concepts, characteristics, and clinical validation of IBM Watson for oncology. *Hanyang Medical Reviews*, *37*(2), 49-60.
27. Bozyiğit, F., & Kılınç, D. (2022). Practices of Natural Language Processing in the Finance Sector. In *The Impact of Artificial Intelligence on Governance, Economics and Finance, Volume 2* (pp. 157-170). Singapore: Springer Nature Singapore.
28. Örpek, Z., Tural, B., & Özmen, S. (2023). Review on the Use of NLP in the Banking Sector. In *Proceedings of International Conference on Engineering, Science and Technology* (p. 154).
29. Shaik, T., Tao, X., Li, Y., Dann, C., McDonald, J., Redmond, P., & Galligan, L. (2022). A review of the trends and challenges in adopting natural language processing methods for education feedback analysis. *Ieee Access*, *10*, 56720-56739.
30. Alam, A. (2023). Harnessing the Power of AI to Create Intelligent Tutoring Systems for Enhanced Classroom Experience and Improved Learning Outcomes. In *Intelligent Communication Technologies and Virtual Mobile Networks* (pp. 571-591). Singapore: Springer Nature Singapore.
31. Fang, Y. Design and Realization of Database System for Judgement Documents Based on Natural Language Processing. *Database*, *6*(1), 192-200.
32. Arruda, A. (2016). An ethical obligation to use artificial intelligence: An examination of the use of artificial intelligence in law and the model rules of professional responsibility. *Am. J. Trial Advoc.*, *40*, 443.

Chapter 2
NLP in Medicine: Enhancing Diagnostics and Patient Care

Aditi Sharma and Akshi Kumar

Abstract Natural Language Processing (NLP) bridges the gap between human language and machine understanding, allowing computers to process and analyse vast amounts of unstructured text data. It has become a transformative tool in the medical field, enabling healthcare professionals, researchers, and institutions to derive actionable insights from vast amounts of unstructured and semi-structured medical data. Medical NLP is profoundly reshaping the landscape of medicine, unlocking the potential for improved patient care, clinical research, and operational efficiency. The advent of social media has revolutionized numerous aspects of human life, from communication to commerce. By empowering patients to stay informed and involved in their own care, NLP technologies help build a patient-centred healthcare ecosystem that emphasizes collaboration and transparency. As NLP technology continues to evolve, it will be essential to remain vigilant about its ethical implications, especially within the realm of patient-centred care. Emerging fields like ethical AI, explainable AI, and responsible AI offer promising approaches for ensuring that NLP applications align with the principles of patient autonomy, privacy, fairness, and transparency. Further research is needed to establish best practices for deploying NLP in ways that enhance care without undermining ethical standards. This chapter presents different applications of natural language processing in healthcare, how it impacts patient care, and what are the different ethical considerations needed to be taken care of.

A. Sharma (✉)
Department of Computer Science Engineering, Thapar Institute of Engineering and Technology, Patiala, Punjab, India
e-mail: Aditi.sharma@thapar.edu

A. Kumar
School of Computing, Goldsmiths, University of London, London, UK
e-mail: Akshi.Kumar@gold.ac.uk

A. Kumar, S. R. Sangwan (eds.), *Transformative Natural Language Processing*,
https://doi.org/10.1007/978-3-031-88988-2_2

Natural Language Processing (NLP) is a transformative field of artificial intelligence (AI) focused on enabling machines to understand, interpret, and respond to human language. Rooted in the intersection of linguistics and computer science, NLP powers a wide range of applications, from virtual assistants and language translation to sentiment analysis and text summarization. NLP bridges the gap between human language and machine understanding, allowing computers to process and analyse vast amounts of unstructured text data. Unlike structured data, which is organized in tables and fields, natural language data is inherently ambiguous, context-dependent, and rich in variability. These characteristics make it challenging for machines to comprehend, requiring advanced algorithms and models to interpret meaning.

As society increasingly relies on digital communication, NLP continues to grow in importance, shaping industries such as healthcare, finance, education, and entertainment. NLP is the most prominent area of research since it can never go out of scope, linguistics is the most common and important way of humans to communicate with each other. Whether it the communication through speaking or writing, language is used, whichever it might be. To create genuinely intelligent systems/ robots understanding of language is a must. NLP is a very broad research area that contains analyzing the text, recognizing it, understanding it, generating the text, and presenting the text among others. The range of applications of NLP is huge.

The history of NLP can be traced back to the 1950s with the advent of early computational linguistics [1]. Milestones such as the introduction of rule-based systems, statistical approaches, and, more recently, deep learning have propelled the field forward. Modern NLP leverages vast datasets, pre-trained language models, and computational power to achieve remarkable accuracy in language understanding and generation. Some of the major NLP applications are:

- Text Analysis and Understanding
- Text Generation and Completion
- Language Translation
- Sentiment and Emotion Analysis
- Information Extraction
- Conversational Chatbots
- Speech Recognition
- Personalization and Recommendation
- Query response
- Text Classification

With invention of text generative or generative AI, the field has transformed a lot, it has eased the work for a lot of fields, yet it's impact in the field of the medicine is not imperative. Healthcare or medical field is one of the most sensitive domains to apply any automated research tool, one wrong decision could be life threatening for a patient. Any use of technology in the healthcare system should be used with certainty that the software will work properly. There have been manly applications of machine learning research in healthcare, many researchers are employing the AI techniques for analysing the reports, and symptoms to identify the possible disease a patient could have or what could be the best treatment plan for the patient could be. Expert systems are also a very useful product for non-medical professionals to understand their symptoms and disease, but it should also be used with caution [2]. In this chapter we are not going to focus on these applications, but the specific one of use of AI tools, that is natural language processing for the healthcare and medicine field.

This chapter explores the applications of NLP in diagnostics and patient care, detailing the methods and techniques employed, key use cases, ethical considerations, and the future of NLP in healthcare. By providing a comprehensive overview of NLP's role in medicine, this chapter aims to highlight its potential in addressing contemporary healthcare challenges while addressing the ethical concerns that accompany this powerful technology.

2.1 NLP in Medicine

Natural Language Processing enables computers to process and analyse human language, making it a crucial tool for managing the vast and complex data generated in healthcare. Medicine produces enormous volumes of unstructured text, including clinical notes, electronic health records (EHRs), medical literature, and patient communications [3]. NLP technologies unlock actionable insights from this data, enhancing patient care, improving operational efficiency, and advancing medical research. By bridging the gap between human language and computational systems, NLP is addressing some of the most pressing challenges in modern healthcare.

One of the most impactful applications of NLP in medicine is in the management of electronic health records. EHRs often contain unstructured text, such as physicians' notes, diagnostic reports, and treatment plans, making them challenge to analyse efficiently. NLP algorithms can extract meaningful information from these records, such as identifying symptoms, diagnoses, and medications, allowing healthcare providers to make faster and more informed decisions. Automated coding systems, powered by NLP, also streamline administrative processes by assigning standardized medical codes to clinical documentation. This not only reduces the workload for medical professionals but also improves billing accuracy and compliance with regulatory standards.

Beyond documentation, NLP enhances clinical decision-making by analysing large datasets to identify patterns and correlations that might not be immediately apparent to clinicians. Clinical decision support systems (CDSS) use NLP to process medical literature, guidelines, and patient data to offer evidence-based recommendations. For example, these systems can alert physicians to potential drug interactions, suggest alternative treatments, or flag critical findings in radiology reports. By integrating NLP-driven insights into clinical workflows, healthcare providers can deliver safer, more personalized care.

Another critical area where NLP is making strides is patient engagement and communication. Virtual health assistants, such as chatbots, leverage NLP to interact with patients, answering their questions, scheduling appointments, and providing reminders for medications or follow-up visits [4, 5]. NLP-powered symptom checkers help patients describe their conditions in natural language, guiding them toward appropriate care. These tools not only empower patients to manage their health more effectively but also reduce the burden on healthcare systems by addressing minor concerns without the need for in-person consultations.

NLP also plays a pivotal role in advancing medical research and public health. Researchers use NLP to mine biomedical literature and clinical trial data for new insights into diseases, treatments, and drug development [6]. NLP can also process data from unconventional sources, such as social media and patient forums, to analyse real-world evidence and public health trends. During pandemics or disease outbreaks, NLP systems quickly extract and synthesize information from global health reports, enabling rapid response and resource allocation.

Despite its transformative potential, NLP in medicine faces unique challenges. Medical language is highly specialized, with complex terminologies, abbreviations, and variations in documentation styles. Ensuring the accuracy and reliability of NLP models requires extensive domain knowledge and annotated data. Privacy and security are also critical concerns, as NLP systems often process sensitive patient information that must comply with regulations like HIPAA and GDPR.

Figure 2.1 showcase this application of NLP in medicine and healthcare. Looking ahead, the future of NLP in medicine is promising. Advances in deep learning and pre-trained models, such as BioBERT and ClinicalBERT, are improving the accuracy of medical NLP systems [7]. Multimodal approaches, which combine text with

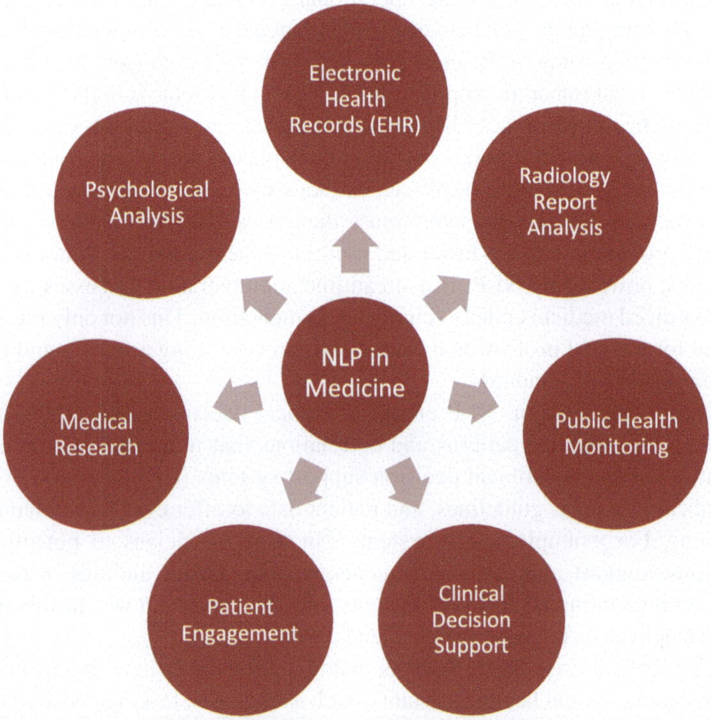

Fig. 2.1 Applications of NLP in medicine

data from imaging, genomics, and wearables, are opening new possibilities for personalized medicine. As these technologies continue to evolve, NLP will play an increasingly central role in delivering better healthcare outcomes, optimizing operations, and driving innovation in medicine.

2.1.1 NLP for Enhancing Patient Care

Beyond diagnostics, NLP is contributing to improved patient care by enabling better communication, increasing access to information, and supporting personalized treatment plans. Several key applications demonstrate how NLP is transforming patient care [2, 5, 8, 9]:

- **Patient Interaction and Communication:** NLP-based chatbots and virtual assistants help patients access health information, schedule appointments, and manage medications. These tools are particularly valuable in managing chronic conditions, where patients benefit from frequent engagement without having to visit a healthcare provider each time. By providing patients with a convenient way to access medical information, NLP reduces healthcare barriers and improves patient satisfaction.
- **Clinical Decision Support:** NLP algorithms can assist clinicians in making evidence-based decisions by retrieving relevant research, treatment guidelines, and clinical trial data based on patient-specific information. These systems can alert clinicians to potential drug interactions, recommend treatment protocols, or identify alternative therapies based on patient data and current medical literature. As a result, clinicians can make more informed and accurate treatment decisions that are tailored to individual patients.
- **Sentiment Analysis for Patient Feedback:** Analysing patient feedback, such as satisfaction surveys and online reviews, can offer healthcare providers insights into the quality of care. NLP-powered sentiment analysis helps healthcare organizations understand patient experiences, identify areas for improvement, and proactively address concerns. By leveraging patient sentiment data, providers can make data-driven changes that enhance the patient experience and improve healthcare outcomes.
- **Mental Health and Emotional Support:** NLP is also making strides in mental health care by analysing language patterns in conversations or written text, helping to detect signs of depression, anxiety, or other mental health conditions. NLP-driven applications can support mental health assessments, offer real-time responses, and even monitor patient well-being over time. These tools are particularly valuable in telemedicine, where healthcare providers may need additional data points to assess a patient's mental state remotely.
- **Predictive Analytics in Patient Outcomes:** Predictive models driven by NLP are being used to anticipate patient outcomes and identify individuals at higher risk of complications. For instance, NLP can analyse past cases and predict fac-

tors that might contribute to hospital readmission or complications, allowing healthcare providers to allocate resources proactively. Predictive analytics in patient outcomes is especially useful in managing high-risk populations, where early intervention can improve patient prognosis.

2.1.2 Role of NLP in Medical Diagnostics

Medical diagnostics often rely on extensive, diverse data sources, including patient history, lab results, medical imaging, and clinical notes. NLP enables healthcare systems to extract relevant information from these sources, particularly unstructured text found in patient records, discharge summaries, and research publications. Here are the keyways in which NLP is revolutionizing medical diagnostics.

- **Analysing Clinical Documentation:** Most patient information exists in electronic health records (EHRs), which contain structured data, such as lab values and unstructured data like physician notes [10]. NLP can parse this unstructured data, identifying crucial medical conditions, past treatments, and comorbidities that aid in diagnostic processes. NLP tools, such as named entity recognition (NER) and sentiment analysis, can automatically extract relevant clinical concepts, thereby minimizing errors in data interpretation and enabling healthcare providers to make quicker, more informed decisions. Healthcare data is diverse, ranging from structured data to unstructured text. NLP algorithms provide a suite of tools for converting this unstructured text into structured, actionable insights. Key NLP techniques used in medical diagnostics include:

  (a) *Named Entity Recognition (NER):* Identifies and categorizes terms within clinical documents, such as symptoms, diagnoses, drugs, and anatomical parts. NER systems are crucial in extracting essential medical information from clinical notes, enabling quick insights into a patient's health status.
  (b) *Sentiment Analysis:* Assesses the emotional tone of patient narratives, which can help identify mental health conditions, patient satisfaction, or distress levels. Sentiment analysis is useful in contexts like monitoring mental health patients or assessing patient experience.
  (c) *Topic Modelling and Clustering:* Groups related information within large datasets, helping healthcare providers and researchers identify patterns and trends in clinical data.
  (d) *Contextual Understanding with Transformers:* State-of-the-art models like BERT (Bidirectional Encoder Representations from Transformers) and its medical adaptations (e.g., BioBERT) enable deep contextual understanding. These models improve the accuracy of diagnosis by extracting nuanced medical knowledge from diverse texts.

By combining these NLP techniques, healthcare providers can generate meaningful insights from unstructured data, significantly enhancing the diagnostic process and enabling the early detection of diseases.

- **Disease Detection and Prediction:** NLP models are increasingly used to detect early signs of diseases by analysing language patterns and specific symptoms documented in clinical records. In the case of chronic diseases like diabetes, cardiovascular disorders, and even some forms of cancer, NLP can detect subtle indicators, which may not be evident to the human eye, to predict disease progression. Early and accurate diagnosis is key to effective treatment, and NLP plays a critical role in analysing patient data to identify early warning signs of diseases. Some examples include:
  (a) *Cancer Detection:* NLP models can analyze pathology reports, imaging studies, and genetic data to identify signs of various cancers early on [11]. For instance, algorithms have been developed to detect patterns in radiology reports that might suggest malignancy, which assists radiologists in identifying potential cases requiring further investigation.
  (b) *Sepsis Prediction:* Sepsis, a life-threatening infection, requires immediate intervention. NLP systems can analyze clinical notes, lab reports, and vital signs to identify high-risk patients and trigger early alerts for healthcare providers.
  (c) *Cardiovascular Disease:* By examining factors such as family history, patient symptoms, and clinical notes, NLP can assist in identifying patients at risk of cardiovascular events, enabling preventative care measures.

NLP-driven diagnostics have demonstrated significant potential in identifying high-risk patients, flagging unusual symptoms, and assisting healthcare providers with timely interventions.

- **Interpreting Radiology and Imaging Reports:** Radiology and imaging reports often contain complex medical jargon, and subtle differences in wording can significantly impact diagnoses. NLP can standardize and interpret these reports, identifying patterns or anomalies that may otherwise be overlooked. By integrating NLP into radiology workflows, healthcare providers can improve diagnostic accuracy, consistency, and speed, particularly in high-stakes areas such as cancer diagnosis and emergency care.
- **Automating Chart Reviews:** Chart reviews an essential part of many healthcare processes often involve manually reviewing patient records to find relevant information. This process is labour-intensive and prone to human error. NLP algorithms can automate chart reviews by summarizing patient history, identifying risk factors, and highlighting essential information, making it easier for clinicians to review cases quickly and accurately. This automation enhances diagnostic efficiency and enables healthcare providers to manage more patients without sacrificing the quality of care.
- **Identifying Adverse Drug Reactions (ADRs):** NLP is instrumental in detecting potential ADRs by analysing clinical notes and patient-reported data. By moni-

toring patterns in text data, such as physician notes and patient complaints, NLP can flag ADRs that may be missed in standard evaluations. Early identification of ADRs is crucial for patient safety, allowing for timely intervention and adjustment of medications to minimize adverse effects.

- **Reducing Documentation Burden:** Physician burnout is a significant issue in modern healthcare, with extensive documentation demands often cited as a major contributor. NLP technologies can significantly reduce the burden of clinical documentation through:

  (a) *Automated Transcription:* NLP can automatically transcribe physician-patient interactions, converting spoken words into structured data without requiring manual data entry.
  (b) *Summarization of Clinical Notes:* NLP algorithms can summarize lengthy clinical notes, providing physicians with concise overviews and key details that enhance their understanding of the patient's condition.
  (c) *Automated Coding and Billing:* NLP tools can assist with assigning diagnostic and procedural codes, reducing the administrative workload on healthcare providers and enabling faster reimbursement.

By automating repetitive tasks, NLP allows providers to focus more on direct patient care and less on documentation, ultimately enhancing patient care quality and provider well-being.

2.2 Sentiment Analysis in Medical Texts

Semantic analysis in the medical field enables a deeper understanding of language by analyzing the meaning of words, phrases, and sentences within medical texts. This process is crucial for extracting valuable insights from vast amounts of unstructured data in electronic health records (EHRs), research papers, clinical notes, and patient reports. Medical texts are highly specialized and often include complex terminologies, acronyms, and context-dependent phrases, which makes semantic analysis both challenging and essential. Semantic analysis in medical NLP can be broadly classified into rule-based techniques, machine learning-based approaches, deep learning models, and hybrid systems that combine these methods. Figure 2.2 shows the most widely used techniques and their unique contributions to semantic analysis.

- **Named Entity Recognition:** NER involves identifying entities by their names and tagging them with categories. For example, "aspirin" could be tagged as a "medication," while "diabetes" might be tagged as a "disease." Rule-based systems use medical dictionaries like UMLS (Unified Medical Language System) to recognize terms, while machine learning-based NER models are often trained on annotated medical datasets to recognize terms and their contexts. Medical NER models need to handle the unique terminologies, abbreviations, and context-

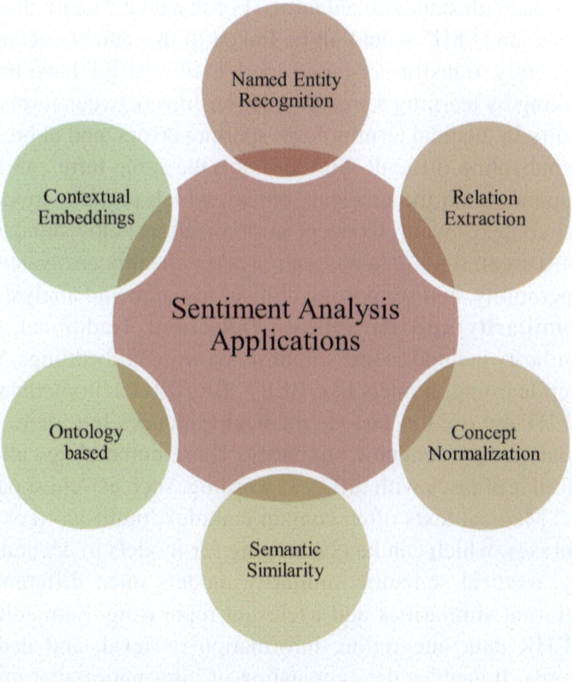

Fig. 2.2 Applications of sentiment analysis in medical texts

dependent meanings in medical texts. For instance, the abbreviation "COPD" could refer to chronic obstructive pulmonary disease or a different concept in non-medical contexts, depending on the text. NER in medical texts is widely used for information extraction, clinical decision support, and structuring EHRs. By identifying key entities in clinical notes, NER facilitates quick information retrieval and aids in summarizing patient records for clinicians.

- **Relation Extraction:** Relation extraction can be approached using rule-based patterns (e.g., using syntactic parsing to capture relationships) or supervised machine learning. In supervised RE, models are trained to recognize specific relationships based on annotated examples. Deep learning models like CNNs (Convolutional Neural Networks) and LSTMs (Long Short-Term Memory networks) are also frequently used to capture more complex relationships within the text. Medical texts often contain indirect or implied relationships, which makes it difficult for models to capture relations that are not explicitly stated. Additionally, the variability in medical language requires RE models to generalize well across different sources. Relation extraction is crucial for building knowledge graphs, detecting drug-drug interactions, and extracting detailed information about disease progression and treatment effectiveness from clinical records.
- **Concept Normalization:** Concept normalization techniques typically combine string matching, rule-based approaches, and machine learning algorithms to

associate entities with standardized terms. For instance, "heart attack," "myocardial infarction," and "MI" would all be linked to the same concept in a medical ontology. Recently, transformer-based models like BERT have been fine-tuned for entity linking by learning semantic relationships between terms and concepts [12]. Variability in medical terminology, spelling errors, and abbreviations make concept normalization difficult. Furthermore, the same term can have different meanings depending on the medical context, which requires models to be both accurate and context aware. Concept normalization helps standardize medical data, which is essential for data integration across healthcare systems, improving EHR interoperability, and supporting clinical research and analysis.

- **Semantic Similarity and Paraphrase Detection:** Traditional approaches to semantic similarity include cosine similarity on word embeddings. With advancements in deep learning, models like BERT, BioBERT (Biomedical BERT), and Sentence-BERT are used to encode medical sentences into dense vector representations that capture semantic similarity. These embeddings allow models to recognize similar phrases with different wording, such as "chest pain" and "pain in the chest." Medical texts often contain complex, multi-word expressions and idiomatic phrases, which can be challenging for models to accurately represent. Additionally, medical semantic similarity models must differentiate between clinically relevant similarities and irrelevant rephrasing. Semantic similarity is applied in EHR data integration, information retrieval, and deduplication of medical records. It enables the aggregation of information that may be phrased differently but carries the same meaning.
- **Sentiment Analysis and Subjectivity Detection:** Sentiment analysis and subjectivity detection in medical texts involve identifying the sentiment and the level of subjectivity within statements. This can be crucial for assessing patient feedback, clinician notes, or monitoring patient emotions. In medical contexts, specialized lexicons for medical sentiment or custom-trained models are often used to capture specific expressions related to patient well-being or clinical outcomes. Medical sentiment analysis is complicated by the clinical tone of medical texts, where negative language does not necessarily imply a negative sentiment (e.g., "no evidence of disease" is positive clinically but negative linguistically). Sentiment analysis is useful for evaluating patient satisfaction in telemedicine, identifying signs of mental health conditions, and summarizing patient experiences in clinical trials.
- **Contextual Embeddings with Transformer Models:** Transformer models use self-attention mechanisms to create contextualized word embeddings that consider the surrounding context. For medical texts, these models are often pre-trained on domain-specific corpora to ensure they understand the nuances of medical language. Fine-tuning these models on specific clinical tasks further enhances their ability to capture domain-specific meanings. Transformer-based models like BERT, BioBERT, and ClinicalBERT have transformed semantic analysis in medicine by creating contextual embeddings that capture the meaning of words in relation to their surrounding text [11, 13]. Transformer models require large datasets and high computational power, which can be a limitation

in medical NLP, where labelled data is scarce. Additionally, handling very large documents like medical histories can be challenging due to model input length limitations. Transformers are widely applied in named entity recognition, relation extraction, concept normalization, and text summarization in the medical domain, achieving state-of-the-art results across tasks.
- **Ontology-Based Approaches:** Ontology-based semantic analysis involves mapping text to concepts within an ontology, often using rules and string-matching techniques. Some models integrate ontologies with machine learning to enhance entity recognition and relationship extraction. Ontology-based approaches utilize structured knowledge bases, such as UMLS, SNOMED CT, and MeSH (Medical Subject Headings), to understand and categorize medical concepts. These ontologies provide semantic relationships between concepts, enhancing the understanding of hierarchical and associative relationships. Ontologies need to be updated regularly to keep up with medical advancements, and the lack of interoperability between different ontologies can limit their effectiveness in cross-institutional data sharing. Ontology-based approaches are used for information retrieval, data integration, and developing knowledge graphs, where understanding complex medical relationships is essential.

2.2.1 Applications of Semantic Analysis in Medical Texts

Semantic analysis in medical texts has numerous applications that support both clinical practice and medical research. Some of the most impactful applications include [2, 14, 15]:

- **Clinical Decision Support Systems (CDSS):** By extracting and interpreting patient data from clinical notes, CDSS provides healthcare providers with insights and recommendations to improve diagnostic accuracy and treatment planning. Semantic analysis in CDSS can identify relevant patient information and integrate it with existing medical knowledge for informed decision-making.
- **Patient Monitoring and Risk Prediction:** Semantic analysis techniques applied to patient records and clinical notes enable the early detection of risk factors and potential adverse events. For instance, identifying phrases related to specific symptoms or risk factors allows NLP systems to flag high-risk patients for closer monitoring.
- **Drug Interaction and Adverse Event Detection:** Semantic analysis techniques can identify potential drug-drug interactions or adverse effects by analyzing patient notes and medical literature. Relation extraction and concept normalization are particularly useful for associating drugs with potential side effects, thereby improving medication safety.
- **Medical Literature Mining:** Researchers and clinicians can leverage semantic analysis to mine medical literature for evidence-based practices, treatment options, and clinical outcomes. By extracting entities and relationships, semantic

analysis tools can quickly summarize relevant findings from vast volumes of research papers.

- **Clinical Documentation Improvement and Summarization:** Semantic analysis tools can enhance clinical documentation by generating concise summaries of patient histories, identifying relevant diagnoses, and improving the clarity of records. Summarization techniques help clinicians access essential information quickly, reducing the time spent on record review.

2.3 Challenges of Using NLP in Medicine

Despite its numerous benefits, implementing NLP in healthcare presents several challenges. One of the primary challenges is the complexity and variability of clinical language, as medical terms often have different meanings depending on context, and abbreviations or acronyms may vary between specialties. Additionally, clinical data is highly sensitive, and patient privacy and data security concerns must be addressed, requiring NLP models to comply with strict healthcare regulations like HIPAA in the United States. Some of these challenges are:

- **Complexity of Medical Language:** Medical terminology is complex, domain-specific, and context-dependent. Different clinicians may use varying terms to describe the same condition, and abbreviations or acronyms can have different meanings in different specialties. For NLP systems to be effective, they must be trained on high-quality, comprehensive datasets that account for the nuances of medical language. Building such datasets is a resource-intensive process, as it requires expert input, data annotation, and continuous model updates.
- **Data Privacy and Security:** Medical data is highly sensitive and subject to strict regulatory standards like HIPAA in the United States and GDPR in Europe. NLP applications must comply with these regulations, ensuring that patient information is anonymized, encrypted, and securely managed. Privacy concerns may also limit access to certain data sources, potentially restricting the scope of NLP applications in healthcare settings.
- **Bias and Generalizability:** NLP models trained on specific datasets may not perform well across different patient populations or healthcare settings, leading to biased or inaccurate outcomes. This is particularly concerning in healthcare, where biased predictions could lead to disparities in diagnosis and treatment. Ensuring diversity in training datasets, continuously testing models, and adjusting for potential biases are essential to prevent unintended consequences and ensure fair treatment.
- **Integration into Clinical Workflows:** To be effective, NLP systems need to integrate smoothly with existing clinical workflows without adding to the workload of healthcare providers. User-friendly interfaces, streamlined processes, and effective training programs are essential to facilitate adoption. Additionally, NLP systems must be able to adapt to different EHR systems and healthcare environ-

ments, requiring customization and compatibility testing to function seamlessly across diverse settings.

- **Uncertainty in Clinical Notes:** Clinical notes, written by healthcare providers to document patient conditions, treatment plans, and observations, often feature complex and nuanced language. A significant challenge for NLP in clinical contexts is handling uncertainty, as medical language frequently conveys ambiguous, tentative, or inconclusive information [16]. Effectively addressing this uncertainty is vital for developing reliable NLP models capable of supporting diagnoses, treatment planning, and predictive analytics. Clinical notes are rich with expressions of uncertainty, as healthcare providers frequently record observations, hypotheses, and differential diagnoses that may not be definitive or fully accurate at the time of documentation.

All these challenges impact the performance of models developed for healthcare, but the prominent impact is of complexity of medical language, that also generates the bias and generalizability, but with the invention of large language models, we can handle these issues, with fine tuning, the models can be better deployed for the healthcare analysis. Another major area of concern in implementation of NLP in Medicine is the uncertainty in clinical notes. in next section we are going to discuss some of the NLP techniques that can help to resolve this issue.

2.4 NLP Techniques for Handling Uncertainty in Clinical Notes

Natural Language Processing is revolutionizing how we interact with and understand the vast ocean of textual data [5]. has become an essential tool in healthcare for extracting meaningful information from clinical notes, which are often rich in unstructured text. Clinical notes written by healthcare providers to record patient conditions, treatment plans, and observations contain complex and nuanced language. One of the main challenges for NLP in clinical contexts is handling uncertainty, as medical language frequently reflects ambiguous, tentative, or uncertain information. Effectively managing this uncertainty is crucial for creating reliable NLP models that can assist in diagnoses, treatment planning, and predictive analytics.

2.4.1 The Nature of Uncertainty in Clinical Notes

Clinical notes are rife with expressions of uncertainty. Healthcare providers often record observations, hypotheses, and differential diagnoses that may or may not be accurate or definitive at the time of writing. This uncertainty can take several forms:

- **Hypotheticals and Suspected Diagnoses:** Physicians may record suspected conditions, noting them with phrases like "possible," "likely," or "rule out." These notations indicate that a diagnosis is not confirmed but rather suggested based on initial observations.
- **Conditional Statements:** In some cases, clinicians may record observations that are dependent on future events, for example, "if the fever persists, could indicate infection." These conditional statements add a layer of ambiguity that NLP models need to understand to avoid misinterpretation.
- **Ambiguous Symptoms and Observations:** Clinical notes may describe symptoms that are difficult to quantify or conclusively categorize, such as "mild discomfort" or "intermittent pain." This language reflects the subjective nature of some medical observations.
- **Linguistic Expressions of Doubt:** Words and phrases such as "appears," "suggestive of," or "not ruled out" are often used to convey doubt, signalling that the information may not be definitive. Identifying these linguistic cues is crucial for NLP models attempting to determine the certainty of clinical statements.

Handling these types of uncertainty is essential for developing NLP systems that provide accurate and actionable insights. Misinterpreting uncertain language as definitive can lead to inappropriate clinical conclusions, which may adversely impact patient care.

2.4.2 NLP Approaches for Handling Uncertainty

Several NLP approaches have been developed to handle uncertainty in clinical notes. These methods are grounded in machine learning, rule-based systems, and probabilistic models, each offering different advantages depending on the complexity and goals of the application.

- **Rule-Based Systems and Heuristics:** One of the earliest and simplest approaches to handling uncertainty is through rule-based systems that identify and categorize uncertain language based on specific keywords and patterns [17]. For example, rule-based systems may use predefined dictionaries of terms like "possible," "probable," "suggests," and "rule out" to flag uncertain statements. Rule-based systems are relatively easy to implement and interpret. They work well in settings where uncertainty is explicitly signalled by keywords, such as phrases that indicate doubt or conditional statements. However, Rule-based systems are often limited by their rigidity; they struggle to capture the contextual nuances of uncertainty and may overlook indirect or implied uncertainty. Additionally, these systems are domain-dependent, meaning they require substantial customization for different types of medical language or specialties.
- **Machine Learning Models:** Machine learning approaches, especially supervised learning methods, offer more flexibility and accuracy in detecting uncertainty compared to rule-based methods. Models can be trained on annotated

clinical datasets where uncertainty is labelled, allowing them to learn patterns that correlate with uncertain language.

(a) *Text Classification Models:* Simple machine learning classifiers, such as Support Vector Machines (SVM) and logistic regression, can be used to categorize sentences or phrases as "certain" or "uncertain." These models require labelled training data to learn distinctions between certain and uncertain statements, but they can generalize well to new data.
(b) *Neural Network Models:* Deep learning models, including recurrent neural networks (RNNs) and transformers like BERT (Bidirectional Encoder Representations from Transformers), have proven highly effective in identifying nuanced language patterns associated with uncertainty. These models can be fine-tuned on clinical notes to recognize both explicit and implicit forms of uncertainty.

- Machine learning models are adaptive and can capture complex patterns of uncertainty that rule-based systems may miss. Neural networks excel at contextual understanding, which is essential for interpreting nuanced expressions of uncertainty. However, Training machine learning models requires large, annotated datasets, which can be challenging to obtain in clinical domains. Additionally, deep learning models are often criticized for being "black boxes," offering little transparency in how they detect uncertainty.
- **Probabilistic and Bayesian Approaches:** Probabilistic approaches, particularly Bayesian models, provide a natural framework for handling uncertainty by estimating the likelihood that a given statement is uncertain based on prior data [18]. Bayesian methods can model both the inherent uncertainty in clinical language and the uncertainty in model predictions.

(a) *Bayesian Neural Networks:* These models extend traditional neural networks by incorporating uncertainty estimates into their predictions. For example, a Bayesian network may output a probability distribution indicating the likelihood that a statement is uncertain rather than a definitive classification.
(b) *Hidden Markov Models (HMMs):* HMMs are also used to model uncertainty by estimating the probability of transitions between certain and uncertain states within a text sequence. For example, in clinical notes, an HMM might recognize that "possible pneumonia" is more likely to be followed by further diagnostic tests than "confirmed pneumonia," based on learned probabilities.

- Probabilistic approaches provide not only predictions but also confidence levels, which are useful for understanding the model's certainty about its own predictions. This is valuable in clinical applications where the implications of uncertain information can be significant. However, Probabilistic models require substantial computational resources and are often more complex to train and interpret. Additionally, they may struggle with the intricate linguistic cues that express uncertainty, especially if not enough relevant training data is available.

- **Hybrid Approaches:** Some of the most effective NLP systems combine rule-based methods with machine learning and probabilistic models to create hybrid systems capable of capturing a wide range of uncertain language. For instance, a hybrid system might use rule-based methods to capture explicit indicators of uncertainty and then use a machine learning model to handle more ambiguous cases. This approach leverages the strengths of each method and helps create a more robust uncertainty detection system. A common application of hybrid systems is in NLP pipelines where rules are applied first to filter obvious cases of uncertainty, followed by a machine learning model to handle the remainder. These systems have been used in applications such as identifying adverse drug reactions or extracting differential diagnoses from clinical notes.
- Hybrid approaches are highly customizable and can address a broader range of uncertainties. They can also mitigate the weaknesses of individual approaches, improving overall accuracy. However, Hybrid systems are more complex to implement and require careful tuning and testing to balance the different components effectively.

2.4.3 Applications and Impact of NLP-Driven Uncertainty Handling in Healthcare

Handling uncertainty in clinical notes has a direct impact on several aspects of healthcare, from diagnostics to treatment planning and patient outcomes [2, 3, 19, 20]:

- **Improved Diagnostic Accuracy**: By accurately identifying uncertain language, NLP models help healthcare providers distinguish between confirmed diagnoses and tentative observations. This clarity enables clinicians to prioritize further testing or monitoring for uncertain cases, improving diagnostic accuracy and reducing unnecessary procedures.
- **Enhanced Clinical Decision Support (CDS)**: CDS systems powered by NLP can now provide healthcare providers with more accurate information by flagging uncertain statements and providing relevant context. For example, an NLP-powered CDS tool might highlight that a certain condition was only "suspected" rather than confirmed, prompting clinicians to take appropriate next steps in diagnosis or treatment.
- **Risk Prediction and Management**: Accurate detection of uncertainty allows NLP models to identify patients at higher risk of complications based on ambiguous symptoms or suspected conditions noted by providers. By flagging these cases, NLP systems can improve patient risk stratification and resource allocation, ultimately enhancing patient safety.
- **Automated Chart Review and Summarization**: NLP systems that account for uncertainty can provide more accurate automated summaries of patient charts,

highlighting both certain and uncertain findings. This can help clinicians quickly review essential information without misinterpreting tentative findings as definitive, saving time and reducing cognitive load.
- **Enhanced Training and Research**: By identifying and categorizing uncertain language in clinical notes, NLP can assist in research and training by providing insights into common sources of diagnostic uncertainty. These insights can then be used to develop training materials that improve clinicians' diagnostic skills and foster better communication of uncertainty in medical records.

As NLP technology continues to evolve, its importance in finance will undoubtedly grow, leading to more sophisticated applications and further changing the landscape of the financial industry. For example, the development of more advanced language models like BERT and GPT-4 is enabling more nuanced and accurate sentiment analysis, leading to better predictions and more effective risk management strategies.

2.5 Ethical Considerations in Medical NLP

The advent of Natural Language Processing (NLP) in medicine has offered promising advancements in diagnostic tools, patient care, and administrative efficiencies. By analysing large amounts of unstructured medical data, such as clinical notes, diagnostic reports, and patient histories, NLP is helping healthcare providers identify patterns, make data-driven decisions, and offer personalized care [21, 22]. However, the implementation of NLP in the medical field brings a host of ethical challenges. Addressing these ethical considerations is critical to ensure that the benefits of medical NLP are realized without compromising patient rights, safety, or fairness.

2.5.1 Patient Privacy and Data Security

One of the most critical ethical concerns in medical NLP is maintaining patient privacy. Medical data used in NLP applications often includes sensitive information from Electronic Health Records (EHRs), which are highly protected under regulations like the Health Insurance Portability and Accountability Act (HIPAA) in the United States, and the General Data Protection Regulation (GDPR) in the European Union. NLP systems need vast amounts of textual data to train algorithms effectively. These datasets can include clinical notes, prescriptions, diagnostic records, and more, which are often rich with personally identifiable information (PII).

Ethical Concerns

(a) *Risk of Reidentification:* Despite de-identification efforts, there is still a risk that NLP models could inadvertently expose patient identities, especially when combined with other available datasets.
(b) *Data Storage and Access:* Ensuring that only authorized personnel have access to sensitive data used in NLP is essential to prevent misuse, data breaches, or unauthorized access.

These concerns can be handled using Data Anonymization and De-Identification techniques like pseudonymization, data masking, and the removal of specific identifiers can reduce the risk of privacy violations. For storage and access Federated Learning and Privacy-Preserving Methods can be used. Federated learning allows NLP models to be trained across decentralized data sources, keeping the data within hospital servers, while differential privacy adds noise to the dataset to prevent reidentification.

2.5.2 Algorithmic Bias and Fairness

NLP models learn from historical medical data, which may contain biases that reflect inequalities in healthcare. These biases can manifest in the form of misdiagnosis, disparities in treatment, and underrepresentation of certain populations, such as ethnic minorities, women, and the elderly [23]. When NLP algorithms are trained on biased data, they risk perpetuating or even amplifying these biases.

Ethical Concerns

(a) *Discriminatory Outcomes:* NLP algorithms may provide unequal predictive accuracy or recommendations for different demographic groups, which can worsen health disparities.
(b) *Bias in Training Data:* The medical datasets used to train NLP models often reflect biases inherent in healthcare systems, including disparities in how certain conditions are diagnosed or treated among different populations.

These concerns can be resolved by following the approaches such as:

- *Diverse and Representative Datasets:* Efforts to collect and use balanced datasets that reflect diverse demographics can help mitigate algorithmic bias. This requires investment in gathering data from underrepresented groups and addressing data quality issues.
- *Bias Audits and Fairness Metrics:* Conducting regular audits for bias and using fairness metrics (e.g., demographic parity, equal opportunity) can help measure and address any disparities in model performance.

- *Algorithmic Transparency:* Ensuring transparency in how NLP models are trained, including which datasets and assumptions were used, can foster accountability and trust among users and patients.

2.5.3 Transparency and Interpretability

Many medical NLP models, particularly deep learning models, operate as "black boxes," meaning that their internal decision-making processes are opaque [24]. This lack of interpretability raises concerns about how decisions are made, especially when NLP tools are used for diagnosis or treatment recommendations, where trust and understanding are crucial for both healthcare providers and patients.

Ethical Concerns

(a) *Lack of Explainability:* If a healthcare provider cannot understand why an NLP model made a specific recommendation or prediction, it can be challenging to trust and rely on the model. Lack of explainability can undermine physician confidence in the tool and lead to errors in patient care.
(b) *Informed Consent:* Patients should be aware when AI or NLP is involved in their diagnosis or treatment decisions. However, if models lack interpretability, it can be difficult to communicate the role of these tools effectively to patients.

These concerns can be resolved by following the approaches such as:

- *Interpretable Models and Explanatory Methods:* NLP models with interpretable architectures, such as attention-based models or those using rule-based methods, can help increase transparency. Techniques like SHAP (Shapley Additive Explanations) or LIME (Local Interpretable Model-Agnostic Explanations) can provide insights into how a model arrived at a decision.
- *Clear Communication with Patients and Providers:* Institutions should provide training for healthcare providers to understand and explain NLP tools' predictions and recommendations, ensuring transparency and informed decision-making.

2.5.4 Consent and Autonomy

Consent and autonomy are central to ethical medical practice. Patients have the right to control how their data is used and to make informed decisions about their care. However, in medical NLP, patient data is often used for model training without direct patient involvement or awareness.

Ethical Concerns

(a) *Informed Consent for Data Use:* Patients may not be fully informed about how their data is used to train NLP models, especially if it is anonymized. This can lead to ethical issues regarding autonomy and patient control over their own information.
(b) *Autonomous Decision-Making:* NLP tools that provide recommendations or diagnoses can sometimes make it difficult for physicians to challenge these suggestions, potentially compromising the physician's ability to make autonomous clinical judgments.

These concerns can be resolved by following the approaches such as:

- *Transparent Data Use Policies*: Hospitals and institutions should clearly communicate with patients about how their data will be used, including for AI model training. Consent procedures should be reviewed to align with ethical principles and regulations.
- *Augmentation, Not Replacement*: NLP tools should be designed to support, rather than replace, clinical judgment. Ensuring that human decision-making remains central to patient care is essential to preserving clinician autonomy and patient trust.

2.5.5 Impact on the Healthcare Workforce

The integration of NLP and AI into healthcare settings can lead to shifts in responsibilities, potentially displacing certain roles or leading to an over-reliance on technology for critical decision-making [25]. These changes can have profound ethical implications for healthcare professionals.

Ethical Concerns

(a) *Job Displacement:* Automation of tasks such as diagnosis coding, documentation, and data entry can lead to concerns about job security for administrative and support staff in healthcare.
(b) *Over-Reliance on AI:* Physicians and healthcare providers may become overly dependent on NLP tools, which could result in diminished critical thinking and diagnostic skills over time.

These concerns can be resolved by following the approaches such as:

- *Education and Training*: Providing healthcare workers with training on how to use NLP tools ethically and effectively can help them integrate these tools as part of a broader skill set.

- *Redefining Roles*: Rather than replacing roles, NLP can redefine them by offloading repetitive tasks, allowing healthcare providers to focus on more complex, human-centred aspects of patient care.

2.5.6 Accountability and Liability

When NLP tools are used in diagnosis, treatment recommendations, or other high-stakes medical decisions, determining accountability in case of errors becomes complex. If an NLP model makes an incorrect recommendation that leads to harm, questions of liability arise, particularly around whether responsibility lies with the model developers, the healthcare providers, or the institution.

Ethical Concerns

(a) *Error Attribution:* Assigning blame when an error occurs due to an NLP tool is challenging, especially if providers acted based on the tool's recommendations in good faith.
(b) *Legal and Regulatory Ambiguities:* The rapid advancement of AI technology often outpaces regulatory frameworks, creating legal uncertainties regarding responsibility and accountability.

These concerns can be resolved by following the approaches such as:

- *Clear Guidelines and Liability Frameworks*: Healthcare organizations and policymakers need to establish clear guidelines that outline accountability and liability, particularly in cases where NLP-driven recommendations are involved in patient care.
- *Continuous Monitoring and Quality Assurance*: Regular monitoring of NLP tools in clinical environments can ensure that errors are identified and corrected, reducing the risk of harm and supporting ethical accountability.

While NLP technology has the potential to revolutionize healthcare by improving diagnostics, personalizing treatment, and optimizing workflows, its ethical implementation is essential to safeguarding patient rights and upholding professional standards. Key ethical considerations, including patient privacy, algorithmic bias, transparency, informed consent, workforce implications, and accountability, must be addressed proactively to ensure that NLP is used responsibly and equitably in healthcare settings. Ethical guidelines, transparent policies, and regulatory oversight will be crucial to fostering trust and maximizing the benefits of medical NLP for all stakeholders.

2.6 Case Studies

Natural Language Processing has proven transformative in healthcare, particularly in enhancing diagnostic processes by enabling data extraction, pattern recognition, and contextual understanding within clinical documentation. Some case studies that showcase how NLP applications have improved diagnosis accuracy, efficiency, and accessibility across diverse healthcare settings are discussed in this section.

2.6.1 Automating Diagnosis Coding with NLP at Mayo Clinic

Accurate ICD (International Classification of Diseases) coding is essential for healthcare billing, insurance reimbursement, and health data reporting. Manual coding is time-consuming, costly, and susceptible to errors, especially given the volume of EHRs processed daily at institutions like Mayo Clinic [26]. NLP provided an efficient solution by automating diagnosis code assignment based on unstructured clinical text. The approach used by Mayo Clinic includes the following steps:

- *Data Extraction:* The NLP model was integrated with Mayo Clinic's EHR system to analyze patient records, including clinical notes, discharge summaries, and diagnostic reports.
- *Entity Recognition and Semantic Mapping:* Using NER, the system identified relevant medical terms and phrases indicative of specific diagnoses. NLP models were trained to map these terms to standardized ICD codes by leveraging databases like UMLS and SNOMED CT.
- *Hierarchical Classification Models:* Mayo Clinic's NLP tool incorporated hierarchical classification techniques to differentiate similar conditions (e.g., distinguishing between different forms of diabetes or hypertension) and assign the most specific ICD codes. Deep learning models, including recurrent neural networks (RNNs), were employed to understand sequence information and capture the context within long clinical documents.
- *Human-in-the-Loop Feedback:* Coders provided feedback on the NLP-generated codes, allowing the system to continuously improve its accuracy through supervised learning.

Mayo Clinic was able to implement their system with better outcomes. The automated system achieved a coding accuracy improvement of 25% over previous manual processes. Coders reported fewer discrepancies, leading to more accurate billing and reporting. By automating the bulk of coding tasks, the NLP model reduced coding time by 60%, freeing up coding staff to focus on complex cases and reducing administrative bottlenecks. The reduction in manual coding labour translated into significant cost savings for the clinic, while simultaneously ensuring that codes were assigned more accurately and promptly.

Mayo Clinic's automated coding system demonstrates how NLP can streamline administrative tasks in healthcare. This automation improved efficiency and accuracy in ICD coding, highlighting NLP's potential to support data-driven decisions, optimize healthcare billing processes, and maintain high-quality health records.

2.6.2 Radiology Report Analysis for Lung Cancer Detection at Stanford

Lung cancer is one of the leading causes of cancer-related deaths worldwide, largely due to late-stage detection. Radiologists are often overburdened, and subtle indications of early-stage lung cancer in radiology reports may be overlooked. Stanford University developed an NLP-powered diagnostic tool to analyze radiology reports and automatically flag possible cases of lung cancer [27, 28]. The approach used by Stanford are as follows:

- *Radiology Data Sources:* The NLP model was trained on a large dataset of radiology reports, annotated by radiologists to identify key terms and phrases linked to lung cancer, such as "nodule," "mass," "ground-glass opacity," and "lesion."
- *Transformer Models:* Stanford used transformer-based models, specifically BioBERT and ClinicalBERT, which are fine-tuned versions of BERT (Bidirectional Encoder Representations from Transformers) pre-trained on biomedical texts. These models created context-aware embeddings that allowed the system to understand the nuances of radiology-specific language.
- *Entity Recognition and Contextual Analysis:* The system performed NER to identify specific findings within the text, such as "3 mm nodule in the left upper lobe." Contextual analysis enabled the NLP model to determine whether terms like "nodule" or "opacity" represented potential cancer indicators or benign findings.
- *Risk Scoring and Flagging:* Based on the NLP output, the system generated a risk score for each report, flagging high-risk cases for follow-up. This score took into account factors such as lesion size, location, and presence of associated findings (e.g., irregular borders or calcifications).

Stanford was able to achieve better accuracy with their approach, along with other benefits. The system flagged early-stage lung cancer cases with over 90% accuracy, identifying cases that may have been missed in traditional radiology reviews. By automatically identifying high-risk cases, the NLP tool allowed radiologists to prioritize critical cases, optimizing their workflow and reducing time spent on routine reviews. The NLP-based prioritization enabled faster diagnosis, shortening the time from scan to diagnosis and allowing for quicker intervention.

This study demonstrates how NLP can enhance radiology analysis by systematically identifying language patterns in reports. Stanford's system not only improved diagnostic accuracy but also provided radiologists with a reliable method to prioritize potential lung cancer cases, leading to earlier interventions and better patient outcomes.

2.6.3 Predicting Psychiatric Conditions Through Patient Records

Diagnosing psychiatric conditions such as depression, anxiety, and bipolar disorder requires comprehensive assessments that often span many visits and interviews. However, early indicators are sometimes documented in patient records through language related to mood, behaviour, or affect. NLP systems can help detect these early indicators and streamline the diagnostic process [29]. The Veterans Health Administration (VHA) implemented an NLP model to analyze clinical notes for language indicative of mental health conditions. The model scanned phrases and keywords related to psychiatric symptoms and behaviours, tagging language patterns that correlated with high-risk conditions.

The NLP system combined sentiment analysis with medical NER, extracting terms like "hopeless," "lethargic," "anxious," and "insomnia." These keywords, along with context derived from sentence structures, were used to assess the likelihood of psychiatric conditions. Additionally, the system accounted for temporal markers, analyzing changes in language patterns over time. The NLP model successfully identified over 85% of at-risk cases, providing mental health professionals with a prompt to assess these patients further. It also helped reduce diagnostic delays for conditions such as depression and PTSD, enabling clinicians to provide interventions earlier in the patient's treatment.

This case study demonstrates how NLP can support mental health diagnosis by capturing language patterns that indicate psychiatric conditions, reducing diagnostic times, and enabling timely therapeutic interventions. Many other researchers have used the NLP to identify depression from different linguistic parameters, especially collected from the social media. The posts along with the pattern such as frequency, timings, and context the mental health of a person can be detected, one such model employed by Dr. Kumar is shown in Fig. 2.3 [30].

2.7 Future Directions

The future of NLP in medicine is promising, driven by advancements in machine learning algorithms, the availability of larger datasets, and growing collaboration between healthcare providers and AI developers. Several areas of growth stand out:

- **Transformer Models and Deep Learning**: Recent advances in transformer-based models like BERT (Bidirectional Encoder Representations from Transformers) and GPT (Generative Pre-trained Transformer) have made NLP

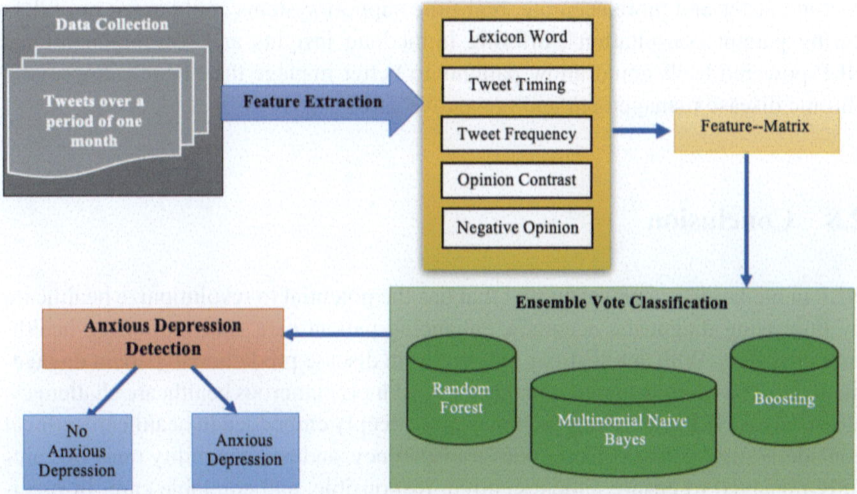

Fig. 2.3 Mental health detection model using social media and NLP

models more accurate, context-aware, and capable of understanding complex medical language. These models can be fine-tuned on specific clinical data, providing precise insights into diagnoses and treatment recommendations.

- **Explainable AI**: As NLP becomes more integrated into healthcare decision-making, there is a need for explainable AI models that provide transparency and accountability. In healthcare, clinicians require clear explanations of how models arrive at their predictions or recommendations. Explainable AI frameworks will enhance trust in NLP applications and encourage wider adoption by ensuring that healthcare providers can confidently interpret and validate AI-driven insights.
- **Interdisciplinary Collaborations**: Future advancements in NLP in healthcare will likely come from closer collaborations between AI researchers, healthcare professionals, and regulatory bodies. Working together, these stakeholders can create high-quality datasets, develop ethical standards, and ensure that NLP systems are both effective and equitable across patient demographics.
- **Integration with Other AI Technologies**: NLP is increasingly being combined with other AI technologies, such as computer vision (for interpreting imaging data) and robotics (for physical assistance in patient care). These integrated systems can provide more comprehensive diagnostic tools, interactive patient care solutions, and advanced support for clinicians. For example, an NLP and computer vision-based system might analyze radiology reports and imaging data together, yielding a more complete diagnostic picture for complex cases.

The potential of NLP in medicine is just beginning to unfold, with promising advancements on the horizon. Combining NLP with genomic data analysis could enable personalized treatments that consider both genetic predispositions and patient histories, significantly advancing precision medicine. As NLP models become faster and more accurate, real-time support systems could assist providers during patient consultations, offering immediate insights and recommendations. NLP-powered tools could allow patients to better manage their health, improving chronic disease management and preventive care.

2.8 Conclusion

NLP in medicine is a powerful tool that has the potential to revolutionize healthcare by improving diagnostic accuracy, enhancing patient care, and supporting healthcare providers. With applications ranging from disease prediction to patient engagement, NLP has demonstrated its capacity to address numerous healthcare challenges. However, as NLP technology becomes more deeply embedded in healthcare, ethical considerations such as privacy, bias, transparency, and accountability must be carefully managed to ensure patient-centred, responsible, and equitable care. In precision medicine, NLP supports by analyzing large volumes of patient data to customize treatments based on individual risk factors, genetic information, and historical treatment outcomes. This enables personalized treatment plans that consider unique patient profiles, advancing both diagnostic accuracy and patient care. Future advancements in NLP for healthcare will continue to push the boundaries of what is possible in diagnostics and patient care, potentially redefining the healthcare landscape.

References

1. Iroju, O. G., & Olaleke, J. O. (2015). A systematic review of natural language processing in healthcare. *International Journal of Information Technology and Computer Science*, *8*(8), 44-50.
2. Roy, K., Debdas, S., Kundu, S., Chouhan, S., Mohanty, S., & Biswas, B. (2021). Application of natural language processing in healthcare. *Computational Intelligence and Healthcare Informatics*, 393-407.
3. Zhou, B., Yang, G., Shi, Z., & Ma, S. (2022). Natural language processing for smart healthcare. *IEEE Reviews in Biomedical Engineering*.
4. Rosruen, N., & Samanchuen, T. (2018, December). Chatbot utilization for medical consultant system. In *2018 3rd technology innovation management and engineering science international conference (TIMES-iCON)* (pp. 1-5). IEEE.
5. King, M. R. (2023). The future of AI in medicine: a perspective from a Chatbot. *Annals of Biomedical Engineering*, *51*(2), 291-295.
6. Sarella, P. N. K., & Mangam, V. T. (2024). AI-driven natural language processing in healthcare: transforming patient-provider communication. *Indian Journal of Pharmacy Practice*, *17*(1).
7. Jain, K., & Prajapati, V. (2021). NLP/deep learning techniques in healthcare for decision making. *Primary Health Care: Open Access*, *11*(3), 373-380.
8. Popowich, F. (2005). Using text mining and natural language processing for health care claims processing. *ACM SIGKDD Explorations Newsletter*, *7*(1), 59-66.

9. Deepti Saraswat, et al. (2022). Explainable AI for healthcare 5.0: opportunities and challenges. *IEEE Access, 10*, 84486-84517.
10. Tan, K. L., Lee, C. P., & Lim, K. M. (2023). A survey of sentiment analysis: Approaches, datasets, and future research. *Applied Sciences, 13*(7), 4550.
11. David Crosby et al. (2022). Early detection of cancer. *Science, 375*(6586), eaay9040.
12. Lavanya, P. M., & Sasikala, E. (2021, May). Deep learning techniques on text classification using Natural language processing (NLP) in social healthcare network: A comprehensive survey. In *2021 3rd international conference on signal processing and communication (ICPSC)* (pp. 603-609). IEEE.
13. Upadhyaya, N., Joshi, H., & Agrawal, C. (2025). Examining NLP for Smarter, Data-Driven Healthcare Solutions. In *Intelligent Systems and IoT Applications in Clinical Health* (pp. 393-420). IGI Global.
14. Baldwin, K. B. (2008). Evaluating healthcare quality using natural language processing. *The Journal for Healthcare Quality (JHQ), 30*(4), 24-29.
15. Locke, S., Bashall, A., Al-Adely, S., Moore, J., Wilson, A., & Kitchen, G. B. (2021). Natural language processing in medicine: a review. *Trends in Anaesthesia and Critical Care, 38*, 4-9.
16. Khattak, W. A., & Rabbi, F. (2023). Ethical considerations and challenges in the deployment of natural language processing systems in healthcare. *International Journal of Applied Health Care Analytics, 8*(5), 17-36.
17. Pressman, S. M., Borna, S., Gomez-Cabello, C. A., Haider, S. A., Haider, C., & Forte, A. J. (2024, April). AI and ethics: a systematic review of the ethical considerations of large language model use in surgery research. In *Healthcare* (Vol. 12, No. 8, p. 825). MDPI.
18. Kumar, A., & Garg, G. (2019). Sentiment analysis of multimodal twitter data. *Multimedia Tools and Applications, 78*, 24103-24119.
19. Kumar, A., & Jaiswal, A. (2020). Systematic literature review of sentiment analysis on Twitter using soft computing techniques. *Concurrency and Computation: Practice and Experience, 32*(1), e5107.
20. Kumar, A., Sharma, A., & Nayyar, A. (2020, February). Fuzzy logic based hybrid model for automatic extractive text summarization. In *Proceedings of the 2020 5th International Conference on Intelligent Information Technology* (pp. 7-15).rep
21. Hripcsak, G., Friedman, C., Alderson, P. O., DuMouchel, W., Johnson, S. B., & Clayton, P. D. (1995). Unlocking clinical data from narrative reports: a study of natural language processing. *Annals of internal medicine, 122*(9), 681-688.
22. Spasic, I., & Nenadic, G. (2020). Clinical text data in machine learning: systematic review. *JMIR medical informatics, 8*(3), e17984.
23. Kumar, A., Sharma, K., & Sharma, A. (2021). Genetically optimized Fuzzy C-means data clustering of IoMT-based biomarkers for fast affective state recognition in intelligent edge analytics. *Applied Soft Computing, 109*, 107525.
24. Ben J Marafino et al. (2018). Validation of prediction models for critical care outcomes using natural language processing of electronic health record data. *JAMA network open, 1*(8), e185097-e185097.
25. Hossain, E., Rana, R., Higgins, N., Soar, J., Barua, P. D., Pisani, A. R., & Turner, K. (2023). Natural language processing in electronic health records in relation to healthcare decision-making: a systematic review. *Computers in biology and medicine, 155*, 106649.
26. Wen, A., Fu, S., Moon, S., El Wazir, M., Rosenbaum, A., Kaggal, V. C., ... & Fan, J. (2019). Desiderata for delivering NLP to accelerate healthcare AI advancement and a Mayo Clinic NLP-as-a-service implementation. *NPJ digital medicine, 2*(1), 130.
27. Aberle, D. R., Gamsu, G., Henschke, C. I., Naidich, D. P., & Swensen, S. J. (2001). A consensus statement of the Society of Thoracic Radiology: screening for lung cancer with helical computed tomography. *Journal of thoracic imaging, 16*(1), 65-68.
28. Kernstine, K. H., Stanford, W., Mullan, B. F., Rossi, N. P., Thompson, B. H., Bushnell, D. L., McLaughlin, K. A., Kern, J. A. (1999). PET, CT, and MRI with Combidex for mediastinal staging in non-small cell lung carcinoma. *The Annals of thoracic surgery, 68*(3), 1022-1028.

29. Levis, M., Westgate, C. L., Gui, J., Watts, B. V., & Shiner, B. (2021). Natural language processing of clinical mental health notes may add predictive value to existing suicide risk models. *Psychological medicine*, *51*(8), 1382-1391.
30. Kumar, A., Sharma, A., & Arora, A. (2019). Anxious depression prediction in real-time social data. *arXiv preprint arXiv:1903.10222*.

Chapter 3
NLP in the Legal Domain: Ensuring Precision and Compliance

K. Sayooj Devadas and Saurabh Raj Sangwan

Abstract This chapter delves into the transformative role of Natural Language Processing (NLP) in the legal domain, focusing on its potential to automate routine tasks, enhance accuracy in contract analysis, and refine litigation strategies. The study underscores the challenges posed by syntactic complexity, domain-specific terminology, and the scarcity of quality data. It highlights the exponential progress of pre-trained language models (PLMs), driven by transformer architectures, in addressing these barriers. The discussion also emphasizes the necessity of Explainable AI (XAI) to ensure transparency and trustworthiness in critical judicial processes. Ethical considerations, including fairness and compliance with legal systems, are critically analyzed, with a focus on democratic nations like India, where precision and unbiasedness in judgments bolster public trust. The chapter concludes with future directions to optimize NLP technologies for efficiency, accessibility, and adherence to legal frameworks, aiming to integrate AI-powered tools seamlessly into judicial workflows while preserving human oversight and comprehension.

K. Sayooj Devadas (✉)
Department of Computer Science and Engineering, Indian Institute of Technology (Indian School of Mines), Dhanbad, Jharkhand, India
e-mail: 24dr0264@iitism.ac.in

S. R. Sangwan
Department of Computer Science and Engineering, Artificial Intelligence and Machine Learning, G L Bajaj Institute of Technology and Management, Greater Noida, Uttar Pradesh, India
e-mail: saurabh.sangwan@glbitm.ac.in

A. Kumar, S. R. Sangwan (eds.), *Transformative Natural Language Processing*,
https://doi.org/10.1007/978-3-031-88988-2_3

The legal domain operates within a complex framework of regulations, statutes, and case law. Precision and compliance are paramount; even minor errors can have significant consequences. Legal professionals grapple with vast volumes of intricate text, from contracts and legislation to court rulings and legal opinions. The sheer complexity and variability of legal language pose challenges for efficient information extraction and analysis.

Natural Language Processing (NLP), a branch of artificial intelligence focused on enabling computers to understand and process human language, offers a powerful toolkit to address these challenges. The legal domain poses unique challenges for natural language processing due to the highly specialized and technical nature of legal language [1]. Nonetheless, there has been significant interest in exploring the potential of NLP technologies to support various legal tasks, from information retrieval and automated contract analysis to litigation outcome prediction and compliance risk assessment [1, 2].

In the academic realm, empirical legal studies have increasingly relied on a variety of computational methods, including NLP, to enhance their analytical capabilities. Meanwhile, commercial enterprises have made attempts to integrate NLP modules into various legal applications, with the goal of streamlining and augmenting legal professionals' workflows [1]. However, the performance of many of these commercial applications has been somewhat inconsistent, as the ability to consistently process legal language with high fidelity remains a significant challenge [1]. This underscores the need for continued research and development in the field of legal NLP, to ensure that these technologies can reliably and precisely handle the intricacies of legal language and ultimately support legal practitioners in their efforts to ensure compliance and mitigate risk.

NLP applications in the legal field range from automating routine tasks like document review to sophisticated analyses such as predicting case outcomes [3, 4]. The evolution of NLP technologies has opened new avenues for enhancing efficiency, accuracy, and compliance in legal practice. For instance, contract analysis, once a laborious and time-consuming process, is now being expedited by NLP-powered tools that can automatically identify key clauses, extract critical information, and even assess potential risks [5, 6]. This not only saves valuable time for legal professionals but also enhances accuracy and consistency in contract review. E-discovery, another area ripe for disruption, has seen the advent of NLP-driven solutions that can sift through mountains of documents to pinpoint relevant evidence, significantly reducing the time and cost associated with litigation [7]. Furthermore, NLP is being leveraged to predict case outcomes by analyzing historical case data and identifying patterns, providing legal practitioners with valuable insights for litigation strategy and settlement negotiations [3]. These are just a few examples of how NLP is transforming the legal landscape, ushering in an era of increased efficiency, accuracy, and compliance.

Recently, Large Language Models (LLMs) have emerged as a transformative force in NLP, demonstrating remarkable capabilities in understanding and generating human-like text. In the legal domain, LLMs hold the potential to revolutionize various aspects of legal practice. Their ability to process and comprehend vast amounts of legal text, coupled with their capacity to generate coherent and contextually relevant responses, opens up new possibilities for automating tasks, improving decision-making, and enhancing access to justice.

Several legal tech companies and institutions are already exploring the use of LLMs. For instance, LLMs are being employed to assist in contract drafting and review, legal research, and even predictive analytics for case outcomes. These implementations aim to streamline legal workflows, reduce costs, and improve the overall efficiency of legal services.

As AI systems, particularly Large Language Models (LLMs), become increasingly integrated into the legal domain, the imperative for transparency and explainability grows exponentially [8]. The "black box" nature of many AI systems, where the internal workings and decision-making processes remain opaque, can breed skepticism and distrust, especially in the legal field where the stakes are high and the implications profound [9]. Whitebox AI, with its emphasis on interpretability and understandability, emerges as a crucial countermeasure to these concerns [10]. By providing clear and comprehensible explanations for the reasoning behind AI-generated outputs, Whitebox AI fosters trust among legal professionals and the public alike [11]. This transparency is not merely a matter of technical elegance; it is fundamental to ensuring the responsible and ethical deployment of AI in the legal domain [12]. When individuals can understand the factors that influenced an AI-powered legal decision, they are more likely to accept and trust the outcome, even if it is not in their favour [13]. This, in turn, can lead to greater acceptance of AI technologies within the legal system, promoting their wider adoption and integration.

This chapter embarks on a journey to explore the dynamic interplay between LLMs and the legal domain, with a particular focus on how these technologies can bolster precision and compliance. We will delve into the core NLP techniques that underpin legal applications, examine real-world case studies that showcase the transformative power of LLMs, and grapple with the ethical and regulatory considerations that accompany the integration of AI into the legal system. By bridging the chasm between cutting-edge technology and the time-honoured traditions of legal practice, we aspire to illuminate a path toward a future where LLMs serve as indispensable allies to legal professionals, enhancing their efficiency, accuracy, and ultimately, their ability to deliver justice.

3.1 Understanding Legal Language and Texts

Legal language is characterized by its formalism, ambiguity, and syntactic complexity [14]. One of the key aspects of this challenge is the highly specialized and technical nature of legal language, which often involves complex sentence structures, domain-specific terminology, and nuanced semantic relationships [9]. Legal texts are rich with specialized jargon, Latin terms, and precise wording designed to avoid ambiguity and ensure clarity in legal interpretation. This makes parsing and understanding such texts a non-trivial task for NLP systems. The use of domain-specific vocabulary, legal jargon, and archaic terms adds further layers of intricacy. Legal texts often feature lengthy sentences with multiple clauses and embedded references, making their interpretation challenging even for seasoned professionals.

Processing legal texts presents unique hurdles for NLP systems. The complexity of sentence structures, extensive cross-referencing, and the potential use of multiple languages within a single document demand sophisticated parsing and analysis techniques. With its centuries-old traditions and emphasis on precision, legal language presents a unique set of challenges for natural language processing (NLP) systems. The intricacies of legal texts stem from several factors:

- **Syntactic Complexity:** The syntactic labyrinth of legal language presents a formidable challenge for NLP systems. Lengthy sentences, often spanning multiple lines or even paragraphs, are commonplace in legal documents. These sentences, laden with numerous clauses, nested phrases, and embedded lists, can confound traditional NLP parsers, making it difficult to accurately identify the core components of the sentence—the subject, verb, and object—and to discern the intricate relationships between its various parts. Moreover, the legal domain is replete with specialized terminology and archaic expressions, a lexicon often impenetrable to general-purpose NLP models. This domain-specific vocabulary can obscure meaning and lead to misinterpretations of syntactic structures and semantic relationships.
- **Legal Jargon and Archaic Language:** The legal domain is notorious for its extensive use of specialized terminology and archaic expressions, often referred to as "legalese." This linguistic peculiarity stems from the historical evolution of legal language and the need for precision and consistency in legal interpretation. However, this very characteristic poses a significant hurdle for natural language processing (NLP) systems.
- **Archaic Expressions:** Legal language often retains archaic words and phrases that have fallen out of common usage, such as "heretofore," "hereinafter," and "witnesseth." These expressions can further confound NLP systems, as they may not be recognized or correctly interpreted.
- **Latin Terms:** The historical influence of Roman law on many legal systems has resulted in the continued use of Latin terms in legal documents, such as "prima facie," "mens rea," and "actus reus." These terms can pose challenges for NLP models that are primarily trained on English text.

To tackle these limitations, researchers are exploring advanced parsing techniques, such as deep learning-based models and graph-based parsing, which can better capture the long-range dependencies and intricate relationships within legal sentences [15]. Considering the limitations and challenges caused by legal texts, the importance of developing specialized NLP models tailored to the legal domain also cannot be ignored [16]. For instance, the development of annotated corpora of legal documents, such as contracts, can provide a crucial foundation for training and evaluating legal NLP systems [16]. NLP offers a range of tools to address these challenges. Tokenization, lemmatization, and part-of-speech tagging help break down complex sentences into manageable units and identify the grammatical roles of words. Parsing algorithms enable the analysis of intricate sentence structures, while Named Entity Recognition (NER) facilitates the identification of key entities like judges, dates, and organizations within legal documents [14].

3.1.1 Ensuring Transparency and Fairness in AI Models

Ensuring transparency and fairness in NLP models is fundamental for creating systems that users can understand and trust. Transparency allows users to see how inputs affect outputs, while fairness prevents biased outcomes that can harm individuals or groups. The following techniques support transparency and fairness in NLP applications:

- **Transparency Techniques:**
  (a) *Explainable AI (XAI):* Explainable AI [17] provides tools for understanding and visualizing how a model arrives at its outputs. Methods like Layer-wise Relevance Propagation (LRP) [8] and Integrated Gradients [9] highlight important words or phrases in input data that contribute to the model's prediction. In an NLP application for medical diagnosis, for example, these tools can help healthcare providers understand which symptoms or keywords led the model to flag a particular condition, allowing them to assess the relevance of the prediction.
  (b) *Attention Mechanisms:* Common in transformer-based models like BERT [10] and GPT [11], attention mechanisms can reveal which parts of an input the model is focusing on. By visualizing attention weights, developers can observe which words or phrases influenced the prediction. In sentiment analysis, attention mechanisms can show how certain words affect the model's interpretation of sentiment, providing users with insight into the model's decision-making process.
  (c) *Interpretable Embeddings:* Word embeddings, which represent words as dense vectors, can be difficult to interpret. Techniques such as Principal Component Analysis (PCA) and t-SNE (t-distributed Stochastic Neighbour Embedding) allow developers to visualize and interpret relationships between words [12]. For example, if a model embeds the word "doctor" closer to "male" than "female," it may indicate an underlying bias. Visualizing embeddings can help developers detect and address these biases.
  (d) *Model Cards and Datasheets for Datasets:* Model cards document a model's design, intended use cases, limitations, and known biases. Datasheets describe a dataset's characteristics, origin, collection methods, and potential biases [13]. By making this information accessible, developers and users can make informed decisions about how to use the model responsibly. For example, a model card might indicate that an NLP model trained primarily on Western media sources may not generalize well to non-Western contexts.

3.2 Core NLP Technologies for Legal Applications

The fusion of diverse NLP techniques, now bolstered by the prowess of Large Language Models (LLMs), is revolutionizing the way machines interact with and interpret legal texts. This synergy enables a deeper comprehension of the intricate nuances and complexities inherent in legal language, empowering legal professionals to navigate the vast and often convoluted landscape of legal information with unprecedented efficiency and accuracy. By dissecting the elaborate sentence structures that characterize legal documents, NLP algorithms, particularly when enhanced by the contextual understanding of LLMs, can accurately identify the core components of a sentence and discern the relationships between its various elements. This ability to parse complex syntax is crucial for extracting meaningful information and ensuring precise interpretation. Furthermore, NLP's capacity to pinpoint key entities within legal texts, such as names of individuals, organizations, locations, and legal concepts, facilitates efficient information retrieval and knowledge organization, streamlining tasks like contract analysis, due diligence, and e-discovery. By uncovering the semantic relationships between these entities, NLP systems can provide a more holistic understanding of the legal context, enabling legal professionals to make more informed decisions.

3.2.1 *Key NLP Technologies and Their Applications*

- **Text Classification and Categorization:** The ability to automatically classify and categorize legal documents is a cornerstone of efficient legal information management and retrieval. NLP algorithms, particularly machine learning models, have proven instrumental in automating this process, enabling legal professionals to swiftly navigate vast repositories of legal texts and pinpoint relevant information. Beyond Support Vector Machines (SVMs) and neural networks, a plethora of machine learning models have been successfully applied to legal text classification tasks.
- **Named Entity Recognition (NER):** NER systems, trained on annotated legal corpora, identify and classify key entities within legal texts, such as names of individuals, organizations, locations, dates, and legal concepts. This facilitates information extraction, knowledge organization, and the automation of various legal tasks, such as contract analysis, due diligence, and e-discovery. Recent advancements in NER models, particularly those leveraging transformer architectures like BERT, have significantly improved the accuracy and efficiency of entity recognition in legal texts [14]. LLMs can further refine NER by leveraging their broader language understanding to disambiguate entities and resolve coreferences, leading to more precise and contextually aware entity recognition.
- For instance, in contract analysis, an NER system can automatically identify and extract crucial information like the names of the contracting parties, the effective

date of the contract, and any specific clauses or provisions mentioned. This streamlines the contract review process, enabling legal professionals to quickly identify key information and potential risks. In the context of e-discovery, NER can be used to identify and classify sensitive information, such as personally identifiable information (PII) or confidential business information, within large volumes of documents, aiding in compliance with data privacy regulations.

- **Semantic Role Labelling (SRL) and Parsing:** Semantic Role Labelling (SRL) transcends the mere identification of entities within legal texts; it delves into the intricate web of relationships between these entities, discerning their roles and functions within the context of a sentence. By unravelling the semantic roles of entities, such as who is performing an action (agent), who or what is being affected (patient), or what tools or means are being used (instrument), SRL provides a more nuanced and comprehensive understanding of legal arguments and contract clauses. This deeper level of analysis empowers a range of legal NLP applications, including automated contract summarization, the identification of obligations and liabilities, and the extraction of legal arguments from case law. Parsing, the process of analysing the grammatical structure of a sentence, complements SRL by providing a structural framework for understanding the relationships between words and phrases.
- **Sentiment Analysis:** Sentiment analysis, the computational process of identifying and categorizing opinions expressed in text, holds immense potential in the legal domain. It can provide valuable insights into judicial attitudes, potential biases, and public sentiment towards legal issues, aiding legal professionals in understanding the emotional landscape surrounding a case or legal matter. However, the application of sentiment analysis to legal texts is not without its challenges. The formal and often neutral language used in legal writing can obscure underlying sentiments, making it difficult for traditional sentiment analysis models to accurately gauge the emotional tone.

- The applications of sentiment analysis in the legal domain are manifold. It can be used to:

  (a) Analyze judicial opinions: Sentiment analysis can help identify the attitudes and biases of judges towards specific legal issues or parties involved in a case, potentially aiding in litigation strategy and case preparation.
  (b) Gauge public sentiment: Sentiment analysis of social media discussions and online forums can provide insights into public opinion on legal matters, helping legal professionals and policymakers understand the social and political implications of their decisions.
  (c) Assess the persuasiveness of legal arguments: Sentiment analysis can be used to evaluate the emotional impact of legal arguments, potentially aiding in the drafting of more persuasive and impactful legal documents.
  (d) Information Retrieval and Extraction: The sheer volume of legal information available can be overwhelming for legal professionals. NLP-powered search engines and information extraction tools enable efficient retrieval of relevant legal documents and specific information from vast databases. These tools

employ techniques like keyword search, semantic search, and question answering to provide targeted and contextually relevant results, significantly accelerating legal research and due diligence processes. LLMs can further enhance information retrieval and extraction by enabling more sophisticated query understanding and providing more nuanced and contextually relevant search results.

(e) Text Summarization: Automated summarization tools condense lengthy legal documents into concise summaries, facilitating quick comprehension and review. This can be particularly valuable for legal professionals who need to quickly assess the key points of a document without reading it in its entirety. Both extractive summarization, which selects and combines the most important sentences from the original text, and abstractive summarization, which generates new sentences that capture the essence of the original text, are being explored for legal text summarization. LLMs, with their generative capabilities, can potentially generate more fluent and informative abstract summaries of legal documents, further aiding legal professionals in their work.

3.3 Leveraging Large Language Models (LLMs) in the Legal Domain: Case Studies and Real-World Applications

The rise of Large Language Models (LLMs) has dramatically expanded the horizons of Natural Language Processing (NLP) across diverse domains, including the legal field. Trained on massive text corpora, LLMs exhibit an impressive capacity for understanding and generating human-like text, enabling them to tackle a wide array of complex tasks. In the legal context, LLMs are proving to be particularly valuable, offering the potential to revolutionize various aspects of legal practice. For instance, LLMs can analyze complex legal documents, such as contracts and legislation, identifying key clauses, extracting relevant information, and even assessing potential risks. This can significantly expedite the contract review process and improve the accuracy of legal due diligence. Moreover, LLMs can be utilized to generate legal documents, such as briefs, pleadings, and contracts, by leveraging their ability to understand legal language and generate contextually appropriate text. This can save legal professionals significant time and effort, allowing them to focus on more strategic tasks. Beyond document analysis and generation, LLMs can also be employed to conduct legal research, providing lawyers with quick access to relevant case law, statutes, and legal arguments. This can significantly streamline the legal research process and enhance the efficiency of legal decision-making. Furthermore, LLMs can assist in predicting case outcomes by analyzing historical case data and identifying patterns, providing lawyers with valuable insights for litigation strategy and settlement negotiations. The ability of LLMs to understand and generate human-like text also opens possibilities for developing AI-powered legal assistants that can answer legal questions, provide legal advice, and even participate in legal negotiations. This can potentially democratize access to legal services and empower individuals to better understand and navigate the legal system.

3.3.1 Contract Drafting and Review

LLMs as Legal Assistants: The drafting and review of contracts are fundamental tasks in legal practice, demanding meticulous attention to detail, precision in language, and a comprehensive understanding of legal principles. The advent of Large Language Models (LLMs) has opened exciting new possibilities for automating and enhancing these tasks, promising to revolutionize the way legal professionals handle contracts.

LLMs, trained on massive datasets of legal texts, can assist in contract drafting by:

- Suggesting Clauses: LLMs can analyze existing contracts and legal databases to suggest relevant clauses and provisions based on the specific needs and circumstances of a contract. This can help ensure that contracts are comprehensive and address all essential legal aspects.
- Identifying Potential Risks: LLMs can identify potential risks and loopholes in contracts by analyzing the language and structure of the agreement. This can help legal professionals mitigate risks and ensure that contracts are legally sound and protect the interests of their clients.
- Ensuring Compliance with Legal Standards: LLMs can be trained on specific legal frameworks and regulations to ensure that contracts comply with relevant legal standards. This can help prevent costly legal disputes and ensure that contracts are enforceable.

3.3.2 Legal Research and Analysis

As discussed in the earlier section, the sheer volume and complexity of legal documents, ranging from statutes and case law to contracts and legal opinions, can make legal research a daunting task. However, the advent of Large Language Models (LLMs) is transforming this landscape, offering unprecedented capabilities for analyzing vast amounts of legal text, extracting relevant information, and summarizing key points, thus significantly accelerating legal research. Recent publications highlight the transformative potential of LLMs in this domain. For instance, a recent study [18] highlighted the efficacy of LLMs in analyzing legal documents and generating comprehensive summaries, allowing legal professionals to quickly grasp the key arguments and precedents relevant to a case. Another study [19] showcased the ability of LLMs to answer complex legal questions with high accuracy, effectively acting as AI-powered legal research assistants.

Recent publications highlight the growing interest and advancements in this area:

- Litigation Outcome Prediction: Similarly, Gray et al. [20] explored the application of LLMs in predicting litigation outcomes, considering factors such as the parties involved, the legal issues at stake, and the jurisdiction of the court.
- Legal Judgment Prediction: A study by Wang et al. [21] investigated the use of LLMs for predicting legal judgments, demonstrating their ability to analyze case facts and legal arguments to forecast the likely outcome of a case.
- Settlement Negotiation Support: LLMs can also assist in settlement negotiations by providing insights into the strengths and weaknesses of each party's case and predicting the likely outcome of a trial, enabling more informed and strategic decision-making.

These studies underscore the transformative potential of LLMs in predictive analytics and decision support for legal tasks. By offering data-driven insights into the likely outcomes of legal cases, LLMs can empower legal professionals to make more informed decisions, develop more effective legal strategies, and ultimately enhance the efficiency and fairness of the legal system.

3.3.3 Compliance and Risk Assessment

The increasing regulatory scrutiny and the growing complexity of legal frameworks make compliance and risk assessment critical tasks for legal professionals. LLMs, with their ability to analyze vast amounts of legal text and understand nuanced legal concepts, are emerging as valuable tools for proactive risk mitigation.

Case studies and current implementations further demonstrate the value of LLMs in compliance and risk assessment:

- LexCheck: This AI-powered contract management platform uses LLMs to automate compliance checks, identifying clauses that violate specific regulations or internal policies.
- Kira Systems: This contract analysis software employs LLMs to identify and assess risks in contracts, providing legal professionals with insights into potential red flags and areas of concern.
- Wolters Kluwer: This legal research and publishing company is exploring the use of LLMs to monitor regulatory changes and provide real-time updates to its legal databases, ensuring that legal professionals have access to the latest legal information.

These examples illustrate the growing adoption of LLMs in compliance and risk assessment tasks, enabling legal professionals to proactively mitigate risks, ensure compliance with legal standards.

One prominent application is in contract analysis and compliance monitoring, where NLP algorithms can automatically review contracts, identify key clauses and provisions, assess potential risks, and ensure compliance with relevant regulations.

For instance, LawGeex, an AI-powered contract review platform, employs NLP to automatically identify and analyze clauses in contracts, flagging potential risks and inconsistencies. This significantly accelerates the contract review process and improves accuracy, saving legal professionals valuable time and resources.

Beyond contract analysis, NLP is also streamlining legal research by enabling efficient retrieval and analysis of relevant case law. NLP-powered tools can sift through vast databases of legal documents, identify precedents relevant to a specific case, and extract key information such as the legal issues involved, the arguments presented, and the court's decision. This empowers legal professionals to quickly access and analyze relevant case law, strengthening their legal arguments and enhancing their litigation strategies. For example, ROSS Intelligence, an AI-powered legal research platform, leverages NLP to analyze case law and provide lawyers with relevant precedents and legal arguments, significantly reducing the time and effort required for legal research.

Another critical application of NLP is in e-discovery and litigation support. The discovery process in litigation, which involves identifying and reviewing potentially relevant documents, can be incredibly time-consuming and expensive. NLP-driven e-discovery solutions automate this process by intelligently classifying and categorizing documents based on their content and relevance to the case. This not only reduces the burden on legal teams but also helps ensure that no critical piece of evidence is overlooked. A case study by Relativity, a leading e-discovery software provider, demonstrated that their NLP-powered platform helped a legal team reduce the time required for document review by 60%, resulting in significant cost savings and improved efficiency.

Furthermore, NLP models are being used to predict the likely outcomes of new cases by analyzing historical case data and identifying patterns. These predictive analytics tools can provide valuable insights to legal professionals, aiding in litigation strategy, settlement negotiations, and risk assessment. A study demonstrated the use of NLP to predict the outcomes of cases before the European Court of Human Rights, showcasing the potential of these technologies to enhance legal decision-making [3]. These advancements in NLP are not only automating mundane tasks but also empowering legal professionals with powerful tools to analyze information, make informed decisions, and ultimately deliver more efficient and effective legal services.

3.4 LLMs for Evidence Management and Analysis

One of the key challenges in the legal domain is the increasing volume and complexity of evidence, which can include text documents, images, audio recordings, and video footage. Traditional methods of evidence management and analysis can be time-consuming and inefficient, requiring legal professionals to manually sift through large amounts of data to identify relevant information. LLMs, with their ability to analyze vast amounts of data and extract relevant insights, can significantly streamline this process, enabling legal professionals to focus on more strategic tasks.

For instance, LLMs can be used to analyze text documents, such as contracts, legal briefs, and court filings, to identify key clauses, extract relevant information, and summarize key arguments. This can help legal professionals quickly understand the relevant facts and legal issues in a case, saving them valuable time and resources.

In addition to text analysis, LLMs can also be used to analyze non-textual data, such as images, audio recordings, and video footage. For example, LLMs can be used to identify and extract relevant evidence from surveillance footage, such as the identification of individuals, objects, and events of interest. This can be particularly useful in criminal investigations, where surveillance footage can provide crucial evidence.

Moreover, LLMs can be used to analyze social media data to identify and extract relevant evidence, such as posts, comments, and images that may be pertinent to a case. This can help legal professionals gather evidence, understand public sentiment, and identify potential witnesses or suspects. However, the use of social media data in legal proceedings raises ethical and privacy concerns that need to be carefully addressed.

- **Surveillance Data Analysis:** In countries like China, where extensive CCTV surveillance networks blanket urban landscapes, the sheer volume of video data generated presents both a challenge and an opportunity for legal proceedings. This abundance of visual information, while potentially invaluable for investigations and prosecutions, can be overwhelming for human analysts to sift through manually. However, the emergence of Large Language Models (LLMs), coupled with advancements in computer vision technologies, offers a promising solution [17, 22]. LLMs, trained on massive datasets of text and code, can be adapted to analyze visual data by associating textual descriptions with corresponding visual features. This enables them to identify and extract relevant evidence from video footage, such as identifying individuals, objects, and events of interest, flagging potentially relevant evidence for further review by legal professionals. For instance, an LLM could be trained to recognize specific actions, like a physical altercation or the exchange of an object, and flag those instances within hours of surveillance footage. This not only accelerates the evidence collection process but also reduces the reliance on manual review, minimizing the risk of human error and bias. Furthermore, by integrating LLMs with facial recognition technology, investigators can potentially identify suspects or witnesses more efficiently, even in crowded or low-resolution footage. However, the deployment of such systems must be accompanied by robust safeguards to ensure data privacy, address potential algorithmic biases, and maintain the interpretability of AI-generated evidence. The ethical implications of using AI to analyze surveillance data, particularly in the context of legal proceedings, must also be carefully considered to ensure fairness and accountability.
- **Social Media Analysis:** Social media platforms have become ubiquitous in modern society, serving as virtual town squares where individuals share their

thoughts, experiences, and perspectives. This abundance of user-generated content, while valuable for communication and social interaction, also presents a unique opportunity for legal professionals. Social media posts, comments, images, and videos can contain a wealth of information relevant to legal proceedings, serving as a valuable source of evidence. LLMs, with their ability to analyze vast amounts of text and multimedia data, are emerging as powerful tools for mining this treasure trove of information. For instance, LLMs can be trained to identify and extract relevant evidence from social media posts, such as those related to a specific case or legal issue. They can also analyze the sentiment and tone of social media discussions to gauge public opinion and identify potential biases.

- **Other Evidence Sources:** Beyond the traditional forms of evidence like documents and witness testimonies, legal professionals now grapple with a deluge of information from diverse sources, including audio recordings, text messages, and emails. These sources, while potentially rich in evidentiary value, often require painstaking manual review, consuming valuable time and resources. However, the emergence of Large Language Models (LLMs) is revolutionizing the way this evidence is managed and analyzed. LLMs, trained on massive text and code datasets, possess the remarkable ability to not only comprehend human language but also to discern patterns, extract key information, and even generate summaries from various data formats. In the context of legal proceedings, this translates to a powerful capability for automating the review and analysis of evidence, freeing up legal professionals to focus on more strategic tasks. For instance, LLMs can analyze audio recordings of depositions or interrogations, transcribing the speech and identifying key statements or contradictions that may be relevant to a case. Similarly, LLMs can sift through troves of emails and text messages, extracting relevant conversations, identifying key individuals, and flagging potentially incriminating or exculpatory information. This automated analysis can significantly expedite the evidence review process, enabling legal teams to build stronger cases and make more informed decisions. Furthermore, by integrating LLMs with other technologies, such as voice recognition and natural language understanding, the potential for evidence analysis expands even further. For instance, LLMs can be used to analyze the sentiment and tone of voice in audio recordings, providing insights into the emotional state of speakers and potentially revealing hidden meanings or intentions. The ability of LLMs to process and analyze diverse forms of evidence is transforming the legal landscape, promising to enhance the efficiency, accuracy, and fairness of legal proceedings.
- The use of Large Language Models (LLMs) for evidence management and analysis is still in its nascent stages, but the potential benefits are already becoming apparent. These powerful AI models, trained on massive datasets of text and code, possess the remarkable ability to sift through vast quantities of information, discern patterns, and extract relevant insights, making them invaluable tools for legal professionals navigating the increasingly complex world of digital evidence. By automating the process of evidence review and extraction, LLMs can

significantly enhance the efficiency of legal proceedings, enabling lawyers and investigators to focus on higher-level tasks that require human expertise and judgment. For instance, LLMs can analyze complex legal documents, such as contracts and legal briefs, identifying key clauses, extracting relevant information, and summarizing key arguments. This can save legal professionals countless hours of manual review, allowing them to quickly grasp the essential elements of a case and make more informed decisions. Moreover, LLMs can be applied to analyze non-textual data, such as audio recordings, video footage, and images, extracting relevant information and identifying patterns that may not be readily apparent to human observers. This can be particularly valuable in criminal investigations, where LLMs can analyze surveillance footage to identify suspects, track their movements, and even detect subtle behavioral cues that may be indicative of guilt or innocence. The ability of LLMs to analyze diverse forms of evidence and provide insightful summaries can significantly enhance the accuracy and efficiency of legal proceedings, ultimately contributing to a more just and equitable legal system. However, it is crucial to ensure that the use of LLMs in this context is guided by ethical considerations and safeguards for data privacy and security. As these technologies continue to evolve, it is essential to establish clear guidelines and protocols to ensure their responsible and ethical deployment in the legal domain.

- **Transformer-based Models:** Transformer models, particularly the Bidirectional Encoder Representations from Transformers (BERT) [14] and its variants, have sparked a revolution in Natural Language Processing (NLP) across diverse domains, including the legal field. These models excel at capturing contextual information and long-range dependencies in text, making them exceptionally well-suited for handling the complex sentence structures and nuanced meanings often found in legal documents. Unlike traditional NLP models that process text sequentially, transformers employ a self-attention mechanism that allows them to weigh the importance of different words in a sentence simultaneously, capturing the relationships between words regardless of their distance from each other. This enables a more nuanced understanding of the text, crucial for deciphering the intricate and often ambiguous language of legal documents.

- For instance, in legal question answering, BERT-based models have demonstrated remarkable accuracy in retrieving relevant information from legal texts and providing precise answers to complex legal queries. This is particularly valuable for legal professionals who need to quickly access specific information from vast legal databases. In contract analysis, transformer models can identify and extract key clauses, provisions, and obligations, streamlining the contract review process and enabling efficient risk assessment [6]. Moreover, these models can be fine-tuned on specific legal domains, such as intellectual property law or tax law, to further enhance their performance on specialized tasks.

Several legal tech companies and research institutions have embraced transformer-based models to develop innovative solutions for the legal profession. For example, Casetext's CARA A.I. leverages transformer models to analyze legal documents and provide relevant case law and insights to legal professionals, significantly accelerating legal research. Similarly, Kira Systems employs transformer-based NLP to automate the review and analysis of contracts, identifying key provisions and extracting relevant data, saving lawyers countless hours of manual review.

Pre-trained language models, along with BERT and its variants, are transforming the legal NLP landscape, offering a diverse toolkit for addressing the challenges of legal text processing and analysis. Their ability to capture contextual information, long-range dependencies, and nuanced meanings makes them particularly well-suited for handling the complexities of legal language. As these models continue to evolve and improve, we can anticipate even more innovative and impactful applications that will further enhance the efficiency, accuracy, and accessibility of legal services.

- **Specialized Legal NLP Toolkits:** Several open-source and commercial NLP toolkits specifically designed for legal text processing have emerged, such as LegalNLP, PyMuPDF, and SpaCy's legal extension. These toolkits provide pre-trained models, annotation tools, and specialized functionalities for tasks like NER, relation extraction, and legal text classification, facilitating the development of legal NLP applications. For instance, LegalNLP, a Python library, offers pre-trained models for legal NER, relation extraction, and document similarity analysis, enabling developers to quickly build legal NLP applications without starting from scratch. This can be particularly useful for tasks such as identifying legal entities in contracts or determining the similarity between two legal documents. PyMuPDF, another Python library, provides tools for extracting text and metadata from PDF documents, which are commonly used in the legal domain. This functionality is essential for preprocessing legal documents and making them amenable to NLP techniques. SpaCy's legal extension offers pre-trained models and functionalities specifically designed for legal text processing, such as identifying legal entities, extracting key information from contracts, and analyzing case law. This extension leverages SpaCy's efficient NLP pipeline and provides specialized components for legal text processing, enabling tasks such as identifying legal citations, extracting key terms and definitions, and analyzing the sentiment of legal arguments. These toolkits, with their specialized functionalities and pre-trained models, are empowering developers and legal professionals to harness the power of NLP for a wide range of legal applications, from automating contract review to conducting legal research and streamlining e-discovery processes.
- **Graph Neural Networks (GNNs):** Graph Neural Networks (GNNs) have emerged as a powerful tool for capturing complex relationships and interdependencies within legal texts, particularly in tasks like statute law analysis and legal reasoning. Legal documents, by their nature, are replete with intricate connections between different sections, clauses, and precedents. GNNs, with their ability to represent data as graphs and learn from the relationships between nodes and edges, are uniquely suited to capture these interdependencies. For instance, in statute law analysis, GNNs can be used to model the relationships between

different legal statutes, identifying overlaps, contradictions, and hierarchies within the legal framework.

These advancements in NLP and the development of specialized models and tools are paving the way for a future where legal professionals can leverage the power of AI to streamline their workflows, enhance their decision-making, and ultimately deliver more efficient and effective legal services.

3.5 Compliance, Ethics, and Regulatory Concerns

The regulatory landscape, with its intricate web of laws and guidelines, presents a significant challenge for organizations striving to maintain compliance. Documents, often the backbone of organizational operations, need to be meticulously assessed to ensure adherence to specific regulations, such as the General Data Protection Regulation (GDPR) or the Health Insurance Portability and Accountability Act (HIPAA). Traditionally, this involved painstaking manual review, a process prone to human error and inefficiency, especially when dealing with voluminous datasets. However, the advent of Natural Language Processing (NLP) has opened up new avenues for automating and streamlining compliance classification, enabling organizations to navigate the complexities of regulatory frameworks with greater accuracy and efficiency.

One of the key challenges in compliance classification is the inherent ambiguity and complexity of regulatory language. Regulations are often drafted in a highly specialized and technical manner, laden with intricate provisions, exceptions, and interpretations that can be difficult even for legal experts to decipher. To address this, researchers are leveraging advanced NLP techniques, such as contextualized word embeddings and deep learning models, to capture the nuances of regulatory language and accurately classify documents. For instance, a study by Hooda et al. [23] explored the use of transformer-based models, fine-tuned on legal and regulatory corpora, to automatically identify and classify documents containing sensitive personal data under GDPR, demonstrating their ability to discern complex patterns and contextual cues in legal text.

The benefits of automating compliance classification are manifold. By automatically identifying and flagging documents that may violate specific regulations, organizations can proactively mitigate risks, reducing the likelihood of penalties and legal disputes. This automation also significantly improves efficiency and productivity, freeing up compliance teams from tedious manual review and allowing them to focus on more strategic tasks. Furthermore, automated systems offer enhanced consistency and accuracy compared to manual approaches, minimizing the risk of errors and inconsistencies. The scalability of these systems is another advantage, enabling them to handle massive volumes of documents, making them ideal for managing compliance across diverse departments and business units. As NLP technologies, particularly LLMs, continue to advance, we can anticipate even more sophisticated and impactful compliance classification systems that will empower organizations to navigate the complex regulatory landscape with confidence and efficiency.

While Large Language Models (LLMs) hold immense potential for revolutionizing various aspects of the legal profession, their application requires careful consideration and mitigation of potential risks.

3.5.1 Interpretability and Explainability

The 'black box' nature of LLMs is a significant concern in the legal domain, where decisions can have far-reaching consequences. The complexity of these models makes it difficult to understand their internal workings and the reasoning behind their outputs. This lack of transparency can lead to distrust and skepticism, especially in high-stakes legal contexts where fairness, accountability, and the ability to explain decisions are paramount.

For instance, if an LLM is used to predict the outcome of a case, legal professionals need to understand the factors that influenced the prediction to ensure fairness and accountability. Without transparency, it is difficult to assess whether the LLM's prediction is based on relevant legal factors or on biases present in the training data. This lack of explainability can hinder the adoption of LLMs in legal settings, where decisions must be justified and explained to ensure public trust and confidence in the legal system. Several approaches are being explored to address the 'black box' nature of LLMs and enhance their interpretability and explainability. These include:

- **Developing Explainable AI (XAI) Techniques:** XAI techniques aim to make the decision-making processes of AI systems more transparent and understandable. This can help legal professionals understand the reasoning behind LLM-generated outputs and identify potential biases or errors.
- **Incorporating Rule-based Systems:** Combining LLMs with rule-based systems can provide a level of transparency and control over the decision-making process. Rule-based systems rely on explicit rules and logic, making their decisions more easily interpretable.
- **Visualizing Decision Boundaries:** Visualizing the decision boundaries of LLMs can help understand how they classify or categorize different inputs. This can reveal potential biases or inconsistencies in the model's decision-making process.
- **Generating Natural Language Explanations:** LLMs can be trained to generate natural language explanations for their outputs, providing insights into their reasoning and decision-making process.

3.5.2 *Bias and Fairness*

The issue of bias in LLMs is a critical concern, particularly in the legal domain where fairness and justice are paramount. LLMs learn from the data they are trained on, and if this data reflects existing biases in the legal system, the models may inadvertently perpetuate or even amplify these biases.

For instance, if an LLM is trained on a dataset of legal cases where certain demographic groups are disproportionately represented as defendants or where specific types of crimes are more heavily prosecuted in certain communities, the model may learn to associate these groups with criminality or predict harsher outcomes for them. This could lead to biased predictions and recommendations, potentially perpetuating systemic inequalities within the legal system.

Furthermore, biases can also manifest in the language used in legal texts. If the training data contains biased language or stereotypes, the LLM may learn to associate certain groups with negative attributes or behaviors, leading to discriminatory outcomes. For example, if legal documents consistently use gendered language or associate certain ethnicities with specific crimes, the LLM may learn to replicate these biases in its own language generation and decision-making. Addressing bias in LLMs requires a multi-faceted approach:

- **Careful Data Curation:** Ensuring that the training data is diverse, representative, and free from bias is crucial. This involves auditing the data for potential biases, collecting data from a variety of sources, and using techniques like data augmentation to mitigate the impact of underrepresentation.
- **Bias Detection and Mitigation:** Developing techniques to detect and mitigate bias in LLMs is an active area of research. This includes methods for identifying biased predictions, analyzing the model's internal representations for bias, and developing algorithms that can debias the model's output.
- **Ethical Considerations and Frameworks:** Establishing ethical guidelines and frameworks for the development and deployment of LLMs in the legal domain is essential. This includes ensuring transparency, accountability, and fairness in the use of these technologies.

By proactively addressing the issue of bias in LLMs, we can harness their potential while mitigating the risk of perpetuating or amplifying existing inequalities in the legal system.

Data Privacy and Security: Legal documents often contain sensitive personal and confidential information. The use of LLMs in the legal domain raises concerns about data privacy and security. It is crucial to ensure that LLMs are used in a way that protects sensitive information and complies with relevant data privacy regulations.

Over-Reliance and Deskilling: The increasing automation of legal tasks through LLMs raises concerns about over-reliance on these technologies and the potential deskilling of legal professionals. It is important to ensure that LLMs are used as tools to augment the capabilities of legal professionals, not replace them entirely.

3.5.3 *Developing Explainable AI (XAI) Techniques*

Explainable AI (XAI) techniques are crucial for making the decision-making processes of AI systems more transparent and understandable. This transparency is particularly important in the legal domain, where decisions can have significant consequences and require clear justification. XAI techniques can help legal professionals understand the reasoning behind LLM-generated outputs and identify potential biases or errors.

Some of the commonly used XAI techniques include:

- **LIME (Local Interpretable Model-Agnostic Explanations):** LIME explains individual predictions by approximating the complex model locally with a simpler, interpretable model, such as a linear model or decision tree. This allows legal professionals to understand which features or factors contributed most to a specific prediction.
- **SHAP (SHapley Additive exPlanations):** SHAP explains predictions by assigning importance values to each feature based on game theory principles. This helps identify the contribution of each feature to the prediction and understand how different features interact to influence the outcome.
- **Attention Mechanisms:** Attention mechanisms, commonly used in transformer-based models, highlight the parts of the input text that the model focused on when making a prediction. This can help legal professionals understand which words or phrases in a legal document were most influential in the model's decision-making process.
- **Rule Extraction:** Rule extraction techniques aim to extract human-readable rules from complex AI models, making their decision-making processes more transparent. This can help legal professionals understand the underlying logic behind the model's predictions and identify potential biases or inconsistencies.

By employing XAI techniques, legal professionals can gain insights into the reasoning behind LLM-generated outputs, identify potential biases or errors, and make more informed decisions. This transparency is crucial for fostering trust in AI systems and ensuring their responsible and ethical deployment in the legal domain.

3.6 Future Directions and Innovations in Legal NLP

The future of NLP in the legal domain is bright. Advancements in machine learning, natural language understanding, and domain-specific model training promise to further enhance the capabilities of NLP systems. The integration of NLP with other emerging technologies, such as blockchain and smart contracts, has the potential to revolutionize legal processes and create new opportunities for innovation. As the legal sector navigates the integration of NLP and other AI-driven technologies, it will be essential to strike a careful balance between leveraging the potential of these tools to streamline legal workflows and ensuring that their application maintains the highest standards of precision and compliance.

Looking forward, Pretrained Language Models (PLMs) can be instrumental in various legal applications, including urban city planning and aiding administrative tasks. By integrating diverse data sources such as social media, CCTV footage, government records, and even satellite imagery, PLMs can enhance the quality and speed of judicial decisions. The ability of PLMs to efficiently search and organize vast amounts of evidence can significantly expedite legal proceedings and improve the accuracy of outcomes.

3.6.1 Emerging Trends in Legal NLP

Advanced Language Models: The advent of powerful language models like GPT-4 and BERT has significantly improved the ability of machines to understand and generate legal text. These models can capture nuanced meanings, long-range dependencies, and contextual information, enabling more accurate and sophisticated legal NLP applications.

Transfer Learning and Domain Adaptation: Transfer learning, which involves leveraging knowledge learned from one domain to improve performance in another, is proving valuable in legal NLP. By fine-tuning models pre-trained on massive general language datasets, researchers can adapt them to the specific nuances of legal language, achieving better results with less training data.

- Integration with Legal Tech and AI Systems
  (a) Intelligent Legal Assistants: NLP is powering the development of intelligent legal assistants that can automate routine tasks, provide legal advice, and assist in document drafting and review. These assistants can significantly enhance the efficiency of legal professionals, allowing them to focus on more strategic and complex tasks.
  (b) Virtual Law Firms: The rise of virtual law firms, powered by AI and NLP, is transforming the delivery of legal services. These firms can provide online legal assistance, automate document processing, and offer personalized legal advice, making legal services more accessible and affordable.
  (c) Innovations in Legal Contract Drafting and Negotiation Tools: NLP is facilitating the development of innovative tools for contract drafting and negotiation. These tools can analyze contracts, identify potential risks, suggest alternative clauses, and even predict the outcome of negotiations, empowering legal professionals to negotiate more effectively and efficiently.
  (d) Collaboration Between Legal Experts and Data Scientists: The successful development and deployment of legal NLP systems require close collaboration between legal experts and data scientists. Legal experts provide domain-specific knowledge and insights, while data scientists bring technical expertise in NLP and machine learning. This interdisciplinary collaboration

is crucial for ensuring that NLP systems are accurate, reliable, and aligned with the specific needs and ethical considerations of the legal domain.

3.6.2 Looking Forward: LLMs and the Future of Legal NLP

The advent of Large Language Models (LLMs) has opened up new horizons for NLP in the legal domain. These models, trained on massive datasets of text and code, possess the ability to understand and generate human-quality legal text, enabling them to tackle a wide range of complex tasks. LLMs can analyze complex legal documents, such as contracts and legislation, identifying key clauses, extracting relevant information, and even assessing potential risks. They can also be utilized to generate legal documents, such as briefs, pleadings, and contracts, by leveraging their ability to understand legal language and generate contextually appropriate text. Beyond document analysis and generation, LLMs can also be employed to conduct legal research, providing lawyers with quick access to relevant case law, statutes, and legal arguments. Furthermore, LLMs can assist in predicting case outcomes by analyzing historical case data and identifying patterns, providing lawyers with valuable insights for litigation strategy and settlement negotiations. The ability of LLMs to understand and generate human-like text also opens up possibilities for developing AI-powered legal assistants that can answer legal questions, provide legal advice, and even participate in legal negotiations. This can potentially democratize access to legal services and empower individuals to better understand and navigate the legal system.

3.7 Conclusion

Natural Language Processing (NLP) is poised to revolutionize the legal landscape, empowering legal professionals to work more efficiently, accurately, and ethically. By automating routine tasks, streamlining research, and providing deeper insights into legal texts, NLP technologies enable lawyers to focus on higher-value activities, such as strategic decision-making and client advocacy. As NLP continues to evolve, its impact on the legal domain will only grow, shaping the future of legal practice and ensuring greater access to justice for all.

The integration of LLMs into the legal domain holds immense promise for further advancements in legal NLP. LLMs, with their ability to generate contextually relevant and coherent text, can potentially automate a wider range of legal tasks, such as drafting legal documents, providing legal advice, and even participating in legal negotiations. However, the responsible and ethical deployment of LLMs in the legal field necessitates a focus on transparency, explainability, and fairness. By addressing these challenges, we can ensure that LLMs serve as valuable tools that augment the capabilities of legal professionals, ultimately promoting a more efficient, accessible, and just legal system.

The journey of NLP in the legal domain is marked by both remarkable achievements and ongoing challenges. From automating contract review and streamlining legal research to predicting case outcomes and ensuring compliance, NLP has already demonstrated its transformative potential. However, the complexities of legal language, the need for explainability and fairness, and the ethical considerations surrounding the use of AI in law continue to drive research and innovation in this field.

While LLMs offer immense potential, it's crucial to recognize that they are not a one-size-fits-all solution for every legal NLP task. Smaller, specialized models, such as those fine-tuned on specific legal domains or tasks, can often provide greater accuracy and efficiency for certain applications. For instance, BERT-based models fine-tuned on contract law have shown remarkable success in tasks like contract clause identification and risk assessment, outperforming more general-purpose LLMs in some cases. These smaller models, with their focused training and reduced complexity, can be more adept at handling specific legal nuances and terminology. Moreover, the limitations of LLMs, such as their tendency to "hallucinate" or generate factually incorrect information, necessitate careful consideration and human oversight. Striking a balance between leveraging the capabilities of LLMs and utilizing smaller, more specialized models, while always maintaining human expertise and judgment at the forefront, is crucial for the responsible and effective deployment of NLP in the legal domain. By striking this balance, we can get the maximum from novel technologies.

One of the key challenges in integrating AI systems, including LLMs, into the legal domain is ensuring transparency and explainability. The "black box" nature of many AI models can lead to distrust and skepticism, especially in legal contexts where decisions can have significant consequences. Explainable AI (XAI), or Whitebox AI, aims to address this by providing insights into the decision-making processes of AI systems, enabling legal professionals to understand the reasoning behind AI-generated outputs and identify potential biases or errors. This transparency is crucial for fostering trust in AI systems and ensuring their responsible and ethical deployment in the legal domain.

The abundance of data generated by surveillance systems, social media platforms, and other sources presents both a challenge and an opportunity for legal professionals. LLMs, with their ability to process vast amounts of information and extract relevant insights, are emerging as valuable tools for evidence management and analysis. In countries with extensive surveillance networks, LLMs can be used to analyze video footage, identify individuals, objects, and events of interest, and flag potentially relevant evidence for further review by legal professionals. Similarly, LLMs can analyze social media data to identify and extract relevant evidence, such as posts, comments, and images that may be pertinent to a case. However, the use of such data in legal proceedings raises ethical and privacy concerns that need to be carefully addressed.

Furthermore, in countries with diverse linguistic landscapes, like India, NLP technologies can play a crucial role in bridging language barriers and promoting inclusivity within the legal system. By enabling the translation and analysis of legal documents in multiple languages, NLP can ensure that all citizens, regardless of their linguistic background, have equal access to legal information and services. This can empower individuals to understand their rights, navigate legal processes, and engage with the legal system more effectively, ultimately contributing to a more just and equitable society.

As NLP technologies continue to evolve, we can anticipate even more groundbreaking applications that will further revolutionize the practice of law. The future of legal NLP is bright, promising to enhance the efficiency, accuracy, and accessibility of legal services, ultimately contributing to a more just and equitable legal system for all. However, it is crucial to address the ethical and regulatory concerns surrounding the use of AI in law, ensuring that these technologies are used responsibly and ethically to promote fairness, transparency, and accountability in the legal system. In this pursuit, it's imperative to safeguard democratic values and prevent legal systems from becoming a monopoly of technocrats or corporations. The increasing reliance on AI and NLP in law should not diminish the role of human judgment, legal expertise, and ethical considerations. Instead, these technologies should empower legal professionals, enhance transparency, and promote greater access to justice for all, ensuring that the legal system remains a pillar of democracy and a guardian of individual rights. This means ensuring that legal professionals are equipped with the knowledge and skills to effectively use and interpret AI-generated outputs, and that they retain ultimate control over legal decision-making. It also means ensuring that AI systems are developed and deployed in a way that is transparent, accountable, and accessible to all, regardless of their technical expertise. By striking this balance, we can harness the transformative potential of NLP while preserving the fundamental values and principles of our legal system.

References

1. Katz, D. M., Bommarito, M. J., & Blackman, J. (2023). A general approach for predicting the behavior of the Supreme Court of the United States. PLoS ONE, 18(3), Article e0282089.
2. Zhong, H., Xiao, C., Tu, C., Zhang, T., Liu, T., & Sun, M. (2019). Neural networks for joint sentence classification and semantic role labeling. Proceedings of the AAAI Conference on Artificial Intelligence, 33(01), 7370-7377.
3. Aletras, N., Tsarapatsanis, D., Preoţiuc-Pietro, D., & Lampos, V. (2016). Predicting judicial decisions of the European Court of Human Rights: a Natural Language Processing perspective. PeerJ Computer Science, 2, e93.
4. Choi, J., & Yoon, K. (2018). A review of evidence retrieval and extraction in legal text. Artificial Intelligence and Law, 26, 343-364.
5. Chalkidis, I., Androutsopoulos, I., & Michos, A. (2020). BERT for Patent Classification. arXiv preprint arXiv:2004.14227.
6. Qu, Z., Liu, T., Li, Z., & Zhong, H. (2022). COSTA: A Large-scale Fine-grained Legal Domain Chinese Corpus for Contract Elements Extraction. arXiv preprint arXiv:2203.05003.

7. Shen, D., & Huang, R. (2020). E-Discovery Based on Keyword Extraction and Text Classification. 2020 IEEE 4th Information Technology, Networking, Electronic and Automation Control Conference (ITNEC), 1503-1507.
8. Samek, W., Wiegand, T., & Müller, K. R. (2017). Explainable artificial intelligence: Understanding, visualizing and interpreting deep learning models. arXiv preprint arXiv:1708.08296.
9. Doshi-Velez, F., & Kim, B. (2017). Towards a rigorous science of interpretable machine learning. arXiv preprint arXiv:1702.08608.
10. Guidotti, R., Monreale, A., Ruggieri, S., Turini, F., Giannotti, F., & Pedreschi, D. (2018, November). A survey of methods for explaining black box models. ACM computing surveys (CSUR), 51(5), 1-42.
11. Adadi, A., & Berrada, M. (2018). Peeking inside the black-box: A survey on explainable artificial intelligence (XAI). IEEE access, 6, 52138-52160.
12. Mittelstadt, B. D., Allo, P., Taddeo, M., Wachter, S., & Floridi, L. (2016). The ethics of algorithms: Mapping the debate. Big Data & Society, 3(2), 2053951716679679.
13. Kim, B., Khanna, R., & Koyejo, O. O. (2016, November). Examples are not enough, learn to criticize! criticism for interpretability. In Advances in neural information processing systems (pp. 2280-2288).
14. Chalkidis, I., Androutsopoulos, I., & Michos, A. (2019). Legal-BERT: The Muppets straight out of Law School. Proceedings of the 57th Annual Meeting of the Association for Computational Linguistics, 4019-4025.
15. Bommasani, R., Hudson, D. A., Adeli, E., Altman, R., Arora, S., von Arx, S., ...& Liang, P. (2021). On the opportunities and risks of foundation models. arXiv preprint arXiv:2108.07258.
16. Funaki, T., Tsujii, J., & Miyao, Y. (2020). Building a legal case corpus for text summarization. Proceedings of the 12th Language Resources and Evaluation Conference, 4494-4503.
17. Nakashima, Y., & Ogawa, T. (2019). Multimodal AI for evidence search in legal domain. Proceedings of the 2019 Conference on Empirical Methods in Natural Language Processing and the 9th International Joint Conference on Natural Language Processing (EMNLP-IJCNLP): System Demonstrations, 111-116.
18. Wang, J., Zhao, H., Yang, Z., Shu, P., Chen, J., Sun, H., Liang, R., Li, S., Shi, P., Ma, L., Liu, Z., Liu, Z., Zhong, T., Zhang, Y., Ma, C., Zhang, X., Zhang, T., Ding, T., Ren, Y., Liu, T., Jiang, X., & Zhang, S. (2023). Legal evaluations and challenges of large language models (arXiv:2411.10137v1). https://doi.org/10.48550/arXiv.2411.10137
19. Qin, W., & Sun, Z. (2023). Exploring the nexus of large language models and legal systems: A short survey (arXiv:2404.00990v1). https://doi.org/10.48550/arXiv.2404.00990
20. Gray, M., Savelka, J., Oliver, W., & Ashley, K. (2023). Using LLMs to discover legal factors (arXiv:2410.07504v1). https://doi.org/10.48550/arXiv.2410.07504
21. Wang, X., Zhang, X., Hoo, V., Shao, Z., & Zhang, X. (2024). LegalReasoner: A multi-stage framework for legal judgment prediction via large language models and knowledge integration. IEEE Access, 12, 166843-166854. https://doi.org/10.1109/ACCESS.2024.3496666
22. Zhang, K., Wang, Y., & Zhang, L. (2020). Deep learning for video surveillance: A survey. Pattern Recognition Letters, 139, 121-134.
23. Hooda, A., Khandelwal, R., Chalasani, P., Fawaz, K., & Jha, S. (2023). PolicyLR: A logic representation for privacy policies (arXiv:2408.14830v1). https://doi.org/10.48550/arXiv.2408.14830

Chapter 4
Introduction to NLP in Finance: Sentiment Analysis and Risk Management

Ravi Ranjan, Kapil Sharma, and Akshi Kumar

Abstract Financial knowledge and understanding have become increasingly important in today's complex world. The ability to manage money effectively is no longer just a personal skill; it's essential for navigating the challenges and opportunities of modern life, but not everyone has the ability to understand and manage personal finances effectively. Financial market has undergone different ages, with the advancement in technologies, many automated tools can help even the financial illiterate people to understand the market dynamics. The advent of social media has revolutionized numerous aspects of human life, from communication to commerce. One of the most profound impacts has been on the dissemination and consumption of information, particularly in the realm of finance. To make a good investment decision, effective market surveillance systems are needed, that can analyze information related to a particular stock like financial documents, earning reports, news, market sentiment etc. Financial markets are driven by information, much of which is textual data like news articles, social media posts, and company filings. Financial experts and enthusiasts can now share their knowledge and insights through blogging platforms, podcasts, and video content. This has made complex financial concepts more accessible to a wider audience, empowering individuals to make informed financial decisions. Natural language processing helps in sentiment analysis, risk management, and economic forecasting by analyzing those textual documents. This chapter presents different applications of natural language processing in financial analysis, along with the recent research work conducted in the field.
Financial literacy has become an increasingly vital skill in today's world. It encompasses a broad range of knowledge, from budgeting and saving to investing and debt management. As the complexity of financial markets and products grows, the

R. Ranjan (✉) · K. Sharma
Department of Information Technology, Delhi Technological University, New Delhi, Delhi, India
e-mail: raviranjan_23phdit502@dtu.ac.in

A. Kumar
School of Computing, Goldsmiths, University of London, London, UK
e-mail: Akshi.Kumar@gold.ac.uk

A. Kumar, S. R. Sangwan (eds.), *Transformative Natural Language Processing*, https://doi.org/10.1007/978-3-031-88988-2_4

importance of financial literacy cannot be overstated. Financial skills empower individuals, who can make informed financial decisions, protect themselves from scams, build financial resilience, contribute to economic growth, and improve their overall quality of life. By understanding basic financial concepts, people can create budgets, set realistic financial goals, and develop strategies to achieve them. This knowledge can also help individuals avoid common financial pitfalls, such as excessive debt, impulsive spending, and scams.

Financial markets, the interconnected networks where buyers and sellers trade financial instruments, play a pivotal role in the global economy. These markets facilitate the exchange of capital, enabling businesses to invest, consumers to borrow, and governments to finance their activities. One of the primary functions of financial markets is to allocate capital efficiently. By matching investors seeking returns with borrowers in need of funds, markets ensure that capital flows to the most productive uses. This process promotes economic growth and development [1].

4.1 Factors Influencing Financial Market

Financial markets are influenced by a variety of factors, including economic indicators, interest rates, political events, and investor sentiment. These factors can cause market prices to fluctuate, creating both opportunities and risks for participants [2]. Potential Factors that impact the Financial Market could be internal or external as shown in Fig. 4.1.

- **Internal Factors**: Information related to the company that influences their price in the financial market could be the Company Performance, Management quality, and product innovation. The quality of a company's management team can significantly impact its stock price. Effective management can lead to better

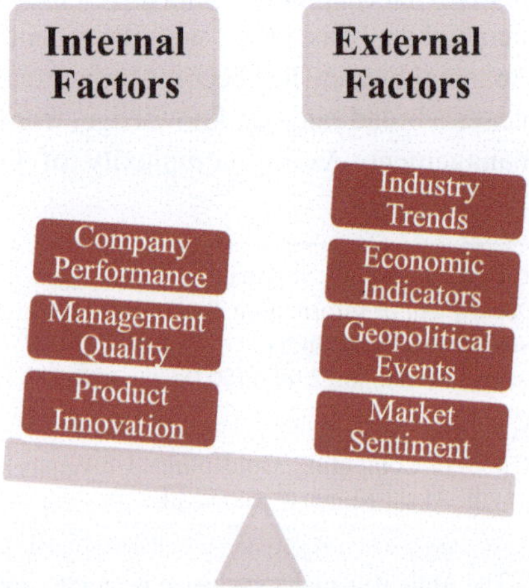

Fig. 4.1 Factors influencing the financial market movements

decision-making, improved operations, and increased shareholder value. Similarly, companies that can innovate and introduce new products or services can gain a competitive advantage and potentially boost their stock prices. The company's performance has the most impact on its dynamics, it could be judged through different parameters like Earnings report, Revenue generated, next year growth prospects. Any company usually publicizes these documents for their stakeholders, the release of these reports changes the investors' sentiment and brings movement into the market.

- **External Factors**: Sometimes a company's market movement has nothing to do with itself, rather it dependent upon the external factors that influence the investors' behavior such as Inflation, Interest rates, GDP, Market sentiment, Geopolitical events, Industry trends among others [2, 3]. A change in any one factor could bring volatility to the market, such as fluctuations in interest rates affecting stock prices. Lower interest rates can make borrowing cheaper for businesses and consumers, which can stimulate economic growth and boost stock prices. Conversely, higher interest rates can slow down economic activity and put downward pressure on stock prices. Global events, such as wars, political instability, or natural disasters, can create uncertainty and volatility in the stock market. The overall mood or sentiment of the market can significantly influence stock prices. Positive sentiment can drive up prices, while negative sentiment can push them down.

4.2 Technological Evolution of Finance

The financial industry has undergone a significant makeover, driven by rapid technological progress. This evolution can be categorized into three distinct eras: the internet age, the era of social media, and the decentralized finance (DeFi) revolution [4]. Each phase has significantly improved how financial information is disseminated, transactions are conducted, and services are delivered.

4.2.1 The Internet Era

The internet boom of the 1990s ushered in a new era of financial information dissemination. Online platforms offered unprecedented access to real-time market data, news, and analysis. Investors can now access financial information from anywhere in the world, at any time. The emergence of electronic trading platforms revolutionized the way transactions were executed, making them faster, more efficient, and more accessible [4]. Moreover, digital communication facilitated seamless interactions between financial institutions, clients, and regulators.

4.2.2 The Age of Social Media

The rise of social media platforms has democratized access to financial information, empowering individuals to make informed decisions about their investments and financial well-being. Prior to the widespread adoption of social media, financial information was primarily controlled by traditional media outlets and financial institutions. This limited access to information and created an uneven playing field, favoring those with the resources to subscribe to premium financial services. However, the emergence of social media platforms like X (previously, Twitter), Facebook, and LinkedIn has challenged this status quo by providing a level playing field for individuals to share and consume financial information.

Social media has played a crucial role in democratizing financial education. This has made complex financial concepts more accessible to a wider audience, empowering individuals to make informed financial decisions. Additionally, social media has facilitated the creation of online communities where people can discuss financial topics, exchange ideas, and learn from each other. Moreover, social media has transformed the way financial information is disseminated. Real-time updates on market movements, economic indicators, and company news are now readily available on social media platforms. This has enabled investors to stay informed and react quickly to changing market conditions. However, the democratization of financial information through social media also presents challenges. The abundance of information available online can be overwhelming, making it difficult for individuals to distinguish between credible sources and misinformation.

4.2.3 The DeFi Revolution

The emergence of decentralized finance (DeFi) marks a pivotal moment in the evolution of the financial industry. This paradigm shift, driven by blockchain technology, is challenging traditional financial institutions and introducing innovative models for managing and exchanging value. DeFi has the potential to transform the way we interact with money, offering greater financial inclusion, transparency, and efficiency.

At the heart of DeFi is blockchain technology, a distributed ledger system that records transactions securely and transparently. Unlike traditional financial systems, which rely on intermediaries like banks, DeFi leverages smart contracts to automate financial processes. These self-executing contracts, written in code, enable peer-to-peer transactions without the need for intermediaries. This eliminates the inefficiencies and costs associated with traditional financial systems. One of the most significant benefits of DeFi is its potential to promote financial inclusion. Traditional financial institutions often exclude individuals and businesses from accessing financial services due to high barriers to entry or lack of trust. DeFi, on the other hand, provides a more accessible and inclusive financial system. However, the DeFi revolution is not without its challenges. Regulatory uncertainties, scalability limitations, and security risks are among the obstacles that need to be addressed. As DeFi continues to evolve, it is essential to develop appropriate regulatory frameworks and address these challenges to ensure its long-term sustainability.

All this information procured during age two and three can be analyzed using automated tools that can analyze and deduce the textual information of such types, so natural language processing is an active area of research for financial market analysis.

4.3 Natural Language Processing in Finance

Natural Language Processing (NLP) is revolutionizing how we interact with and understand the vast ocean of textual data [5]. This revolutionary field is particularly impactful in finance, where words hold immense power in shaping market trends and influencing investor behavior. Financial markets are driven by information, and a significant portion of this information comes in the form of text—news articles, social media posts, company filings, analyst reports, and more. The exponential growth of these textual data sources, coupled with advancements in computational power and algorithms, has made NLP a critical tool for financial professionals [6, 7]. NLP techniques enable the extraction of meaningful insights from this sea of text, transforming unstructured data into actionable intelligence.

Sentiment analysis, a core application of NLP in finance, focuses on understanding the emotions and opinions expressed in financial text. By gauging the sentiment—whether positive, negative, or neutral—associated with specific assets, companies, or market events, analysts can gain valuable insights into market trends and investor behavior. The implications of sentiment analysis extend far beyond understanding market mood. It plays a vital role in risk management, a critical aspect of financial decision-making. Traditional risk management approaches often rely on quantitative data and historical trends [8–10]. However, these methods may fall short in capturing the unpredictable nature of market sentiment, which can quickly shift based on news, events, or even rumors. By incorporating sentiment data into risk models, financial institutions can obtain a more comprehensive view of potential risks and make more informed decisions. Sentiment analysis empowers risk managers to identify and assess various types of risks. For instance, by monitoring news and social media sentiment surrounding a particular company, credit risk analysts can gain insights into its financial health and potential for default [7, 11]. Similarly, market risk managers can use sentimental data to identify potential market bubbles or crashes, allowing for timely adjustments to portfolio allocations. The integration of sentiment analysis with traditional risk management techniques represents a significant step towards a more robust and proactive approach to risk mitigation.

4.4 Applications for NLP in Finance

Natural Language Processing (NLP) is rapidly changing the financial industry, going far beyond just analyzing sentiment. NLP is being applied in diverse areas, from portfolio management and risk assessment to regulatory compliance, by extracting insights from the vast amounts of text data generated in the financial world [12]. Some of the major areas in finance where NLP makes a huge impact by automating the analysis of different financial documents, social media post and news analysis are shown in Fig. 4.2.

- **Predicting market movements**: Advanced NLP techniques are used with machine learning algorithms to forecast stock prices, cryptocurrency trends and other market shifts. These models analyze huge amounts of text from sources like news articles, social media posts and financial reports to identify patterns and predict future trends. Sophisticated models like CNNs are used to process textual data, extracting features and identifying nuanced relationships between words and phrases that can indicate changes in market sentiment.
- **Algorithmic trading**: Hedge funds and other investment firms are increasingly using NLP in their algorithmic trading systems. These systems analyze real-time news and social media data to identify trading opportunities and make trades

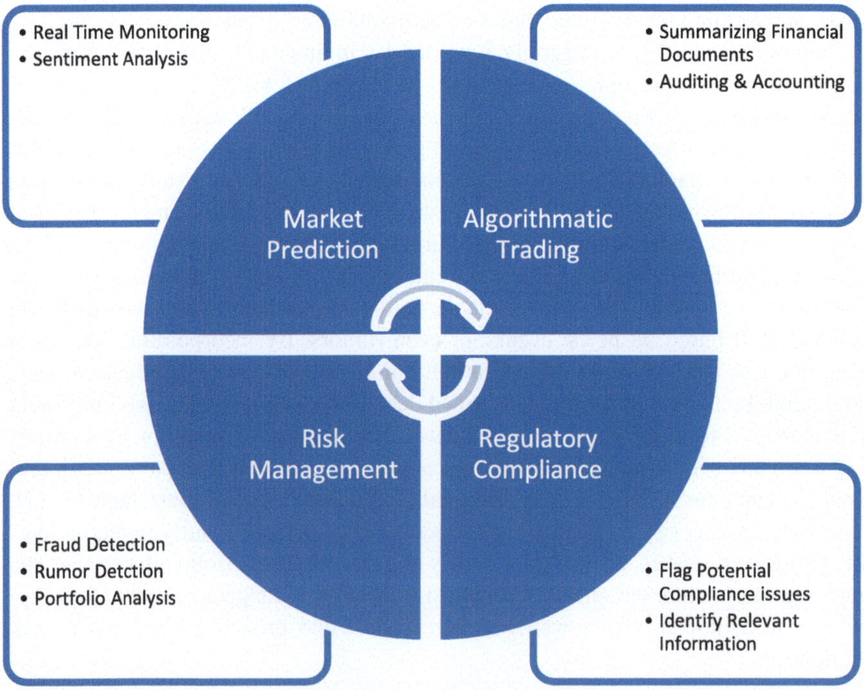

Fig. 4.2 Applications of NLP in finance

automatically. This allows investors to take advantage of the speed and efficiency of NLP to gain an advantage in the fast-paced world of financial markets.
- **Risk management**: Financial institutions are using NLP to identify and assess various types of risks by analyzing textual data from sources like news articles, regulatory filings and internal communications. This allows risk managers to spot potential problems early on and take steps to reduce potential losses. NLP can also be used to monitor social media for sentiment around specific companies or industries to assess potential reputational risks. By understanding public perception, institutions can manage their brand and reduce the risk of financial losses due to reputational damage.
- **Regulatory compliance**: which is vital in the highly regulated financial industry. Financial institutions are using NLP to automate the review of documents, identify potential compliance violations and generate reports. This not only saves time and money but also minimizes the chance of human error.

As NLP technology continues to evolve, its importance in finance will undoubtedly grow, leading to more sophisticated applications and further changing the landscape of the financial industry. For example, the development of more advanced language models like BERT and GPT-4 is enabling more nuanced and accurate sentiment analysis, leading to better predictions and more effective risk management strategies.

4.5 NLP for Understanding Market Sentiment

Sentiment analysis aims to understand the underlying sentiment expressed by experts and investors in their communication/social media posts. By analyzing the language used in text, sentiment analysis can determine whether the sentiment is positive, negative, or neutral [10, 13, 14]. This technique has found widespread applications in various areas, including customer service, brand monitoring, and market research. In the financial world, sentiment analysis plays a crucial role in understanding market sentiment, which can provide valuable insights for investors and traders. The ability to accurately forecast economic trends has always been a coveted goal for investors, policymakers, and businesses alike [15]. Traditional forecasting methods primarily relied on historical numerical data like price fluctuations and economic indicators. However, the rise of digital communication, particularly social media, has unveiled a vast repository of textual data that offers a new dimension to economic forecasting—sentiment.

Market sentiment refers to the overall attitude of investors toward a particular asset or market. It is influenced by a variety of factors, including economic indicators, company news, and geopolitical events [3, 16]. Understanding market sentiment can be challenging, as it is often intangible and can shift rapidly. News articles, social media posts, and financial reports all contain valuable information about how individuals and institutions perceive economic developments. NLP techniques allow us to tap into this wealth of textual data and extract sentiment signals that can potentially enhance traditional forecasting models. Sentiment analysis offers a powerful tool for gaining insights into market sentiment by analyzing the language used in news articles, social media posts, and other text-based sources [17–19].

One of the primary applications of sentiment analysis in finance is to gauge investor sentiment towards specific stocks or industries. By analyzing the sentiment expressed in news articles and social media discussions, analysts can identify emerging trends and potential investment opportunities. For example, a surge in positive sentiment towards a particular company may indicate that it is poised for growth, while a decline in sentiment may suggest that it is facing challenges. Sentiment analysis can also be used to assess the overall market sentiment. By analyzing the sentiment expressed in a variety of sources, including news articles, financial blogs, and social media, analysts can determine whether investors are generally optimistic or pessimistic about the market. This information can be valuable for making investment decisions, as it can help investors anticipate potential market movements. The process involved in Financial Sentiment analysis are discussed next.

4.6 Data Collection and Preprocessing in Financial Sentiment Analysis

The foundation of any successful sentiment analysis project lies in the quality and quantity of the data used. In the context of financial sentiment analysis, this involves collecting relevant textual data and preparing it for analysis. This section will delve into the key steps involved in data collection and preprocessing [20].

4.6.1 Data Collection

Financial sentiment analysis typically relies on a variety of textual data sources, including:

- **News articles:** Financial news outlets provide a wealth of information on market trends, company performance, and economic indicators.
- **Social media:** Platforms like X (previously Twitter), Reddit, and StockTwits offer real-time insights into investor sentiment and discussions.
- **Financial blogs and forums:** Online communities and forums dedicated to finance and investing can provide valuable perspectives and opinions.
- **Financial reports and filings:** Annual reports, quarterly earnings, and other financial disclosures can offer insights into company sentiment and performance.

Some of the prominent data sources that different researchers have focused on for financial sentiment analysis are discussed in Table 4.1.

Table 4.1 Publicly available dataset for financial sentiment analysis

Dataset name	Size	Description
StockNet [21]	88 NASDAQ Stocks	Comprehensive dataset for stock movement prediction from tweets and historical stock prices
Loughran-McDonald Master Dictionary [22]	Include words from 10-K filings and earnings calls	Word lists tailored for financial documents, particularly 10-K filings, identifying sentiment within specific financial contexts
Financial PhraseBank [23]	4840 sentences	Collection of phrases or sentences from financial news texts and company press releases
SemEval-2017 Task 5 [24]	2510 Microblogs and 1647 News	Dataset was created as part of the SemEval-2017 Task 5: Fine-Grained Sentiment Analysis on Financial Microblogs and News
StockTwits [25]	36,787,719	Social media dataset from StockTwits annotated for sentiment, offering real-time analysis capabilities for market prediction
FinTextQA [26]	1262 Question Answer Pair	Dataset for long-form financial question answering (LFQA). Includes a variety of question types like Concept Explanation, Numerical Calc
FiQA Sentiment dataset [27]	~5000 sentences	Annotated financial news and microblogs, focusing on both sentiment and relevance, useful for comprehensive sentiment analysis
FinBERT Dataset [28]	~500,000 articles	BERT-based language model that is pre-trained on a large corpus of financial communication text

4.6.2 Data Extraction and Cleaning

Once the data sources have been identified, the next step is to extract the relevant text data. This can be done using web scraping techniques or APIs provided by the data sources. The extracted data may contain noise, such as HTML tags, punctuation, and stop words. Therefore, it is essential to clean the data to ensure that it is suitable for analysis. Common data cleaning techniques include:

- **Tokenization:** Breaking down the text into individual words or tokens.
- **Stemming:** Reducing words to their root form (e.g., “running” becomes “run”).
- **Lemmatization:** Reducing words to their base form, considering grammatical rules (e.g., “better” becomes “good”).
- **Stop word removal:** Removing common words that do not carry significant meaning (e.g., “the,” “and”, “a”).
- **Noise reduction:** Removing irrelevant or noisy information, such as HTML tags and punctuation.

4.6.3 Data Annotation

To train a sentiment analysis model, the collected and cleaned data must be annotated with sentiment labels. This involves manually labelling each piece of text as positive, negative, or neutral. While this process can be time-consuming, it is essential for creating a high-quality training dataset.

4.6.4 Data Balancing

In many cases, the collected data may be imbalanced, with a disproportionate number of positive or negative examples. This can bias the trained model and lead to inaccurate predictions. To address this issue, various techniques can be used to balance the dataset, such as oversampling minority classes or under sampling majority classes.

Following this process, carefully selecting data sources, cleaning the data, and annotating it with appropriate labels, researchers can create a high-quality dataset that is suitable for training and evaluating sentiment analysis models. This foundation is essential for accurate and reliable sentiment analysis in the financial domain.

4.7 Techniques in Financial Sentiment Analysis

Financial sentiment analysis, at its core, aims to decipher the emotions and opinions embedded within financial texts, providing valuable insights into market trends and investor behavior. The sources provided explore a range of techniques employed in this domain, from lexicon-based methods to sophisticated machine learning and deep learning models, highlighting their strengths, limitations, and potential applications.

4.7.1 Lexicon-Based

Lexicon-based approaches, grounded in pre-defined dictionaries (lexicons) that assign sentiment scores to words and phrases, offer a rule-based foundation for sentiment analysis [29–31]. These scores, aggregated, unveil the overall sentiment of a given text.

- **VADER:** Designed explicitly for the nuances of social media text, VADER (Valence Aware Dictionary for Sentiment Reasoning) stands out as a popular lexicon-based method [29, 30]. VADER surpasses simple polarity (positive/negative/neutral) by factoring in elements like punctuation, capitalization, and intensifiers to gauge sentiment intensity.

- **Domain-Specific Lexicons:** The financial domain, rich in specialized jargon, benefits significantly from lexicons tailored to its unique linguistic landscape. These lexicons capture the nuanced meanings of words in financial contexts, enhancing the accuracy of sentiment analysis [31]. For instance, the Loughran–McDonald lexicon, crafted explicitly for financial texts, proves effective in distinguishing between words with similar meanings but different financial implications [22].

4.7.2 Machine Learning

Supervised learning techniques are being used to learn patterns and relationships between textual features and sentiment labels from labelled training data. Some of the commonly used techniques are:

- **Naive Bayes:** A probabilistic classifier known for its simplicity and efficiency, particularly effective for text classification tasks [32].
- **Support Vector Machines (SVMs):** Powerful classifiers adept at identifying the optimal hyperplane to segregate data points based on sentiment classes [33].
- **Random Forest:** An ensemble learning method combining multiple decision trees to bolster accuracy and robustness, mitigating the risk of overfitting [34].

A cornerstone of machine learning-based sentiment analysis, feature engineering transforms textual data into numerical representations suitable for algorithms [31, 35]. Common features include:

- Bag-of-Words: Represents text as a vector of word frequencies, overlooking word order but capturing the overall vocabulary usage.
- N-grams: Considers sequences of *n* consecutive words, incorporating local word order information for a more context-aware representation.
- TF-IDF: Weighs words based on their document frequency and corpus rarity, highlighting terms crucial to a document's meaning while diminishing the impact of common words.

4.7.3 Deep Learning

Deep learning, a sophisticated branch of machine learning, leverages multi-layered artificial neural networks to extract intricate features from data [35, 36]. These models, excelling in various NLP tasks, including sentiment analysis, have shown remarkable capabilities in capturing complex relationships within text.

- **Recurrent Neural Networks (RNNs):** Well-suited for sequential data like text, RNNs possess a "memory" that allows them to consider the order of words and phrases, making them ideal for capturing the flow and context of sentences.
- **Long Short-Term Memory (LSTM) Networks:** A specialized type of RNN, LSTMs address the "vanishing gradient" problem, enhancing their ability to learn from long sequences. They excel at identifying long-range dependencies in text, crucial for understanding complex financial narratives.
- **Convolutional Neural Networks (CNNs):** Originally designed for image processing, CNNs have proven effective in sentiment analysis by extracting local patterns and features from text, identifying key phrases and sentiment-bearing expressions.

4.7.4 Transformers

Revolutionizing the field of NLP, transformer-based models like BERT (Bidirectional Encoder Representations from Transformers) have achieved state-of-the-art results in sentiment analysis [28, 37, 38]. Their strength lies in capturing long-range dependencies and contextual relationships, deciphering the subtleties of financial language:

- **Fine-Tuning Pre-Trained Models:** Pre-trained on massive text datasets, transformer models can be fine-tuned for specific domains like finance. This adaptation to financial language nuances enhances their accuracy in sentiment analysis tasks.
- **Domain-Specific Variants:** Specialized variants of BERT, such as FinBERT, are explicitly trained on financial texts, further amplifying their ability to capture financial sentiment [28].

4.7.5 Hybrid Approaches

Hybrid approaches, as the name suggests, leverage the advantages of both lexicon-based and machine learning methods. For instance, they might use a lexicon-based approach for initial sentiment scoring, subsequently refined by a machine learning model incorporating contextual information and other features [39].

4.7.6 Beyond Polarity

While sentiment analysis traditionally focuses on positive/negative/neutral classification, some research ventures into a broader emotional spectrum relevant to financial markets. Models identifying emotions like fear, anger, sadness, joy, trust, and anticipation offer a richer understanding of market sentiment. These nuanced emotions can provide deeper insights into investor behavior and potential market shifts [40].

Some of the popular techniques used by different researchers for financial market analysis are presented in Table 4.2.

Table 4.2 Financial sentiment analysis techniques

Model/technique	Approach	Description
FinBERT	Transformer	FinBERT is a variant of BERT pre-trained on financial news data and fine-tuned for financial sentiment analysis
RoBERTa	Transformer	RoBERTa, like BERT, is a transformer-based model optimized with more robust training methods and longer sequences
XLNet	Transformer	XLNet improves upon BERT by using permutation-based training, capturing bidirectional context without masked tokens
VADER	Lexicon	VADER is a lexicon and rule-based sentiment analysis tool tuned for social media
DistilBERT	Transformer	DistilBERT is a lighter, faster version of BERT. It's used in financial sentiment analysis for faster, real-time sentiment classification tasks
GRU	Deep Learning	GRUs, like LSTMs, are recurrent neural networks that capture sequential information but with a simpler architecture and fewer parameters
LSTM	Deep Learning	LSTMs are useful in financial sentiment analysis for processing sequential data, such as time-stamped news articles or tweets, and learning long-term dependencies within the text
GPT (Generative Pre-trained Transformer)	Autoregressive transformer	GPT models, especially GPT-3, are capable of generating coherent text, making them suitable for financial sentiment classification
T5 (Text-To-Text Transfer Transformer)	Transformer	T5 converts every NLP task into a text-to-text problem, making it versatile for financial sentiment analysis

4.8 Assessing the Effectiveness of Financial Sentiment Analysis

Evaluation metrics are crucial for measuring the performance and effectiveness of financial sentiment analysis models. These metrics provide insights into how well a model can capture and quantify sentiment from textual data and its ability to predict market trends. Several evaluation metrics are commonly used in this domain, each offering unique perspectives on the model's capabilities.

One of the most widely used metrics is accuracy, which measures the proportion of correctly classified sentiments (positive, negative, or neutral) compared to the total number of classifications [21, 22]. However, accuracy can be misleading when dealing with imbalanced datasets where one sentiment class dominates. For instance, during a bull market, the majority of news articles might express positive sentiment, leading to high accuracy even for a naive model that simply classifies everything as positive.

Precision and recall offer a more nuanced evaluation, especially with imbalanced datasets. Precision measures the proportion of correctly predicted positive sentiments out of all predicted positive sentiments. A high precision indicates a low rate of false positives. Conversely, recall measures the proportion of correctly predicted positive sentiments out of all actual positive sentiments. High recall suggests that the model effectively identifies the most positive sentiments. The choice between prioritizing precision or recall depends on the specific application. For example, if the goal is to identify a small number of highly probable investment opportunities, precision would be more important. However, if the goal is to capture a broad range of positive sentiments to gauge overall market optimism, recall would be more relevant.

To address the limitations of accuracy with imbalanced datasets, metrics like weighted Macro F1, Area Under Curve (AUC), Receiver Operating Characteristic (ROC), and Matthews Correlation Coefficient (MCC) are often used [25, 29]. These metrics provide a more balanced evaluation by considering the performance across all sentiment classes, even when the classes are unevenly distributed.

When the objective of the sentiment analysis is to predict continuous values, such as stock prices, regression metrics are employed. Common examples include Root Mean Squared Error (RMSE), Mean Squared Error (MSE), Mean Absolute Error (MAE), and hit ratio. RMSE and MSE penalize larger errors more heavily, while MAE provides a more linear assessment of errors. The hit ratio measures the proportion of correctly predicted price movements (up or down).

Beyond these general metrics, some studies focus on the profitability of trading strategies derived from sentiment analysis models. Metrics like cumulative profit, Sharpe ratio, and Maximum Drawdown (MDD) are used to assess the financial performance of such strategies [41]. Cumulative profit reflects the total profit generated over a specific period. The Sharpe ratio measures the risk-adjusted return of a strategy, while MDD quantifies the largest potential loss that an investor might experience.

The selection of appropriate evaluation metrics depends on the specific objectives and the nature of the financial sentiment analysis task. Choosing metrics that align with the application's goals ensures a meaningful assessment of the model's performance and provides valuable insights for further development and refinement.

4.9 Challenges in Financial Sentiment Analysis

While financial sentiment analysis offers valuable insights into market trends and investor behavior, it is essential to acknowledge the inherent challenges and limitations that impede its accuracy and effectiveness [41]. These hurdles stem from the complex nature of financial language, the rapid evolution of online communication, and the inherent subjectivity of sentiment itself.

4.9.1 Domain Specificity

One prominent challenge is the domain specificity of financial language. Unlike general-purpose sentiment analysis models, those employed in finance must be adept at interpreting jargon, acronyms, and technical terminology unique to the financial domain. Failing to grasp these nuances can lead to misinterpretations and inaccurate sentiment classifications. For example, a lexicon-based approach might misinterpret the term "short" in a financial context, confusing it with its negative connotation in general language. To address this, researchers have developed specialized lexicons, such as the Loughran-McDonald dictionary, and fine-tuned language models like FinBERT, which are trained specifically on financial text data.

4.9.2 Sarcastic Reactions

Another significant limitation lies in the difficulty of handling ambiguous and sarcastic language, particularly prevalent in social media posts. Sarcasm and irony often rely on contextual cues and subtle linguistic nuances that machine learning models struggle to interpret accurately. Consider a tweet stating, "Great earnings report! Time to short the stock." A naive sentiment analysis model might misinterpret this as positive sentiment, failing to grasp the sarcastic intent. Developing more sophisticated algorithms capable of detecting such nuances is an active area of research.

4.9.3 *Data Scarcity*

The reliance on annotated data for training supervised machine learning models presents another obstacle. Manually labelling large datasets for sentiment analysis is a time-consuming and resource-intensive process, which can limit the scope of analysis to specific domains or languages. Furthermore, data quality concerns, particularly with social media data, can affect the reliability of sentiment analysis results. Social media posts often contain noise, irrelevant information, and slang that can hinder accurate sentiment extraction.

4.9.4 *Emerging Trends*

The dynamic nature of language and the emergence of new terms and expressions pose ongoing challenges for financial sentiment analysis. Models trained in historical data might struggle to accurately interpret emerging slang or financial neologisms. For instance, a model trained before the advent of cryptocurrencies might misinterpret terms like "HODL" or "FOMO," leading to inaccurate sentiment classification. Continuous model adaptation and retraining are crucial to address this evolving landscape.

4.9.5 *Prior Experience*

The inherent subjectivity of sentiment itself presents a fundamental limitation. Different individuals might interpret the same text with varying emotional responses. Factors like personal experiences, cultural backgrounds, and investment biases can influence sentimental perception. Consider a news article reporting on a company's merger. A risk-averse investor might perceive this as negative, fearing increased market volatility, while a growth-oriented investor might view it positively, anticipating expansion and increased profits. Accounting for this inherent subjectivity remains a significant challenge for financial sentiment analysis.

4.9.6 *Real-Time Analysis*

The speed and virality of social media content require real-time monitoring. Market surveillance systems now incorporate AI-driven tools to track sentiment on platforms like Twitter, Reddit, and even YouTube.

Addressing these challenges and limitations is crucial for advancing the field of financial sentiment analysis. Future research directions include developing more sophisticated algorithms capable of handling nuanced language, incorporating real-time sentiment data into trading platforms, and exploring multi-modal analysis that combines text, images, and other data sources for a more comprehensive understanding of market sentiment. As the sources indicate, the pursuit of more robust, accurate, and context-aware sentiment analysis models will continue to drive innovation and improve the effectiveness of this valuable tool in the financial domain.

4.10 NLP for Financial Compliance and Risk Assessment

The financial landscape has undergone a seismic shift in recent decades, spurred by rapid technological advancements, increasing global interconnectedness, and a surge in regulatory scrutiny following the 2008 financial crisis. Financial institutions today face mounting pressure to navigate a complex web of compliance requirements while safeguarding against evolving financial risks. Amid this challenging backdrop, artificial intelligence (AI) has emerged as a transformative force, offering innovative solutions to streamline compliance processes, bolster risk management, and enhance overall financial stability. NLP can be used for analyzing manipulations in market to avoid the fraud activities, and it could help in ensuring regulatory institutions for compliance with market regulations.

4.11 NLP in Detecting Financial Anomalies

Market manipulation on social media poses significant challenges for financial market surveillance due to the vast volume and speed of information flow, the anonymity of users, and the difficulty in distinguishing between genuine market sentiment and false information. Social media allows manipulators to rapidly spread misleading content across multiple platforms, making it hard for surveillance systems to keep up and for regulators to trace the origins of manipulative activities [42, 43]. The informal and often ironic language used online further complicates sentiment analysis, while varying legal frameworks across jurisdictions create additional enforcement barriers. These factors necessitate more advanced surveillance technologies and updated regulatory approaches to effectively address social media-driven market manipulation.

Financial market surveillance refers to the systematic monitoring of financial markets to detect and prevent manipulative, abusive, or illegal practices. The primary goal is to ensure transparency, protect investors, and maintain the integrity of the markets. Regulatory bodies such as the Securities and Exchange Board of India (SEBI), U.S. Securities and Exchange Commission (SEC), Financial Conduct Authority (FCA), and others worldwide implement and oversee surveillance mechanisms to uphold market fairness and maintain public confidence in the financial system.

Surveillance involves monitoring trading activities, investigating irregularities, and ensuring that the market complies with set standards and regulations [10, 43]. It has become increasingly complex due to globalization, technological advancements, and the growth of automated and algorithmic trading systems. NLP can be used for financial market surveillance, especially using social media monitoring, we can analyse different types of activities that could lead to identification of manipulative tasks, such as.

- **Prevention of Market Manipulation:** Market manipulation refers to attempts by individuals or groups to artificially influence the price or volume of a security for personal gain. Preventing market manipulation is one of the primary goals of financial market surveillance. By preventing manipulation, surveillance systems ensure that the market remains fair and that prices reflect genuine supply and demand forces, protecting the interests of honest investors.
- **Avoiding Fraudulent Activities:** Fraudulent activities in financial markets include illegal practices such as insider trading, accounting fraud, and the dissemination of false or misleading information to influence stock prices. Avoiding fraud is crucial to maintaining market confidence and protecting investors. Avoiding fraudulent activities ensures investor protection and market transparency, preventing significant financial losses caused by illegal schemes or misinformation.
- **Recognizing Unfair Trading Practices:** Unfair trading practices include a range of activities that give certain market participants an unfair advantage, such as front-running (trading based on non-public information about an upcoming large transaction) and wash trading (creating the illusion of market activity by repeatedly buying and selling the same asset). Recognizing and stopping unfair trading practices maintains a level playing field in financial markets, ensuring that all participants operate under the same rules and standards.

Financial market surveillance plays a pivotal role in maintaining the stability, fairness, and transparency of the global financial system. With the increasing complexity of financial markets due to globalization, technological innovations, and the proliferation of social media, the need for advanced and adaptive surveillance techniques is more important than ever [44, 45]. By employing AI-driven systems, cross-market cooperation, and real-time monitoring of social media, regulators can continue to safeguard the integrity of financial markets and protect investors from fraud and manipulation.

4.12 Social Media Analysis for Risk Assessment

In recent years, social media has become an influential force in shaping global financial markets. Platforms like Twitter, Reddit, and Facebook, among others, have grown into spaces where individuals and groups share market opinions, strategies, and even insider information. These platforms have evolved from mere communication tools into powerful drivers of investor sentiment and market movements. As social media's influence has grown, so has the need for financial market surveillance to monitor these activities to protect market integrity and prevent manipulation. Given the influential role of social media, regulators and financial institutions have started incorporating social media analysis as part of broader financial market surveillance frameworks. Monitoring online discussions can provide early warnings of market manipulation, fraudulent activities, or coordinated market actions that could destabilize the financial system.

4.12.1 *Information Dissemination and Sentiment Shifts*

Social media has democratized access to financial information, allowing retail investors to voice opinions and share insights that can shape market movements. What was once the domain of financial analysts and institutional investors is now open to the masses. Social media platforms have become places where financial news, rumours, and speculative analysis spread rapidly, affecting investor sentiment and market dynamics.

The GameStop short squeeze in January 2021 demonstrated the sheer power of retail investors on platforms like Reddit's WallStreetBets. A large group of individual investors drove up the stock price of GameStop, resulting in massive losses for institutional investors who had shorted the stock. This event showed how social media can quickly generate collective actions that impact stock prices and market trends.

4.12.2 *Rapid Market Movements*

Social media has made financial markets more reactive and volatile. A single tweet, post, or rumour can lead to immediate and significant market shifts. For example, tweets from influential figures like Elon Musk can cause the price of stocks or cryptocurrencies to spike or crash within moments.

- **Elon Musk and Bitcoin:** Musk's tweets about Bitcoin and Dogecoin have caused rapid price fluctuations, showcasing the power of social media in driving market behaviour as shown in Fig. 4.3. His tweets caused Bitcoin prices to surge or fall, influencing retail and institutional investors alike. Mr. Musk posted a tweet mentioning his company and DogeCoin in April 2021, the price of the cryptocurrency reached its all-time high value. The surge in price has no underlying internal factor, but only the external factor of support of a famous person.
- **Market Behavior and Herding Effect:** Social-Media often amplifies the herding effect, where investors follow trends based on crowd sentiment. When a large

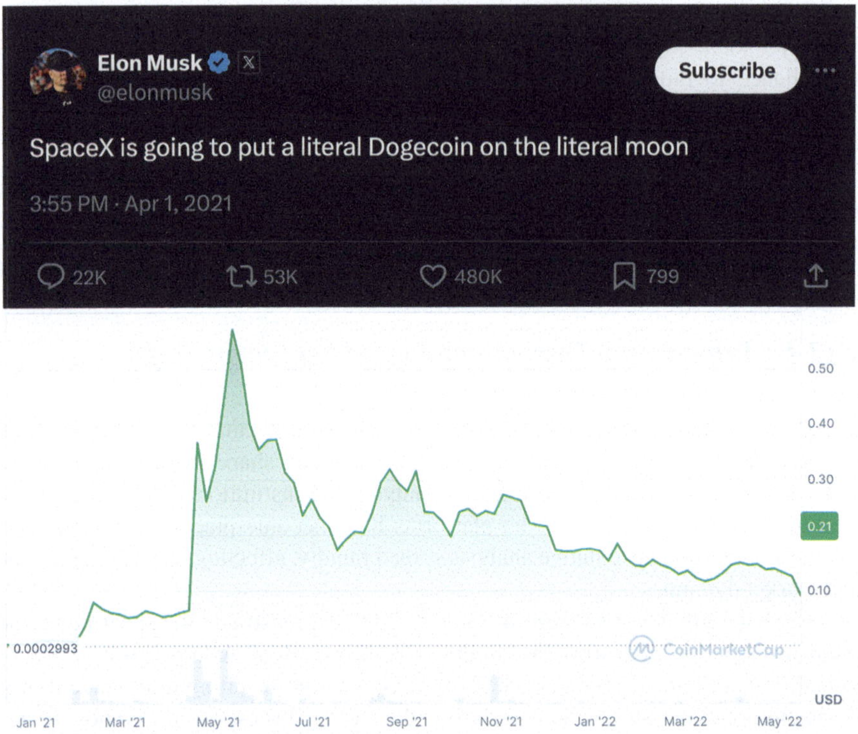

Fig. 4.3 Rapid market movement

group of investors acts in concert, driven by social media hype, it can lead to exaggerated market reactions.

- **Detecting Market Manipulation**: Market manipulation, such as pump-and-dump schemes or coordinated short squeezes, is increasingly organized or influenced by discussions on social media. Financial surveillance tools must now extend to these platforms to catch signs of illegal behaviour before they lead to large-scale market disruptions.

  (a) *Pump-and-Dump Schemes:* Social media platforms can be used to spread misleading information about a stock or cryptocurrency, leading to a sharp increase in price (pump) followed by a coordinated sell-off (dump) that leaves unsuspecting investors at a loss. Surveillance systems analyse posts and discussions to detect the early stages of such schemes.

  (b) *Coordinated Short Squeezes*: Social-Media makes it easier for groups of retail investors to coordinate actions like short squeezes. By monitoring social media platforms, regulators can identify unusual collective actions and pre-emptively address market risks.

The speed and virality of social media content require real-time monitoring. Market surveillance systems now incorporate AI-driven tools to track sentiment on platforms like Twitter, Reddit, and even YouTube. These tools use NLP to detect potential market-moving events, identifying keywords, hashtags, or trends that could signal coordinated trading actions or manipulative behaviour.

Social media has become a powerful force in the financial markets, influencing stock prices, investor sentiment, and market trends. As its role continues to grow, regulators and financial institutions must adapt by integrating social media surveillance into their broader market oversight frameworks. Real-time monitoring of social media discussions, sentiment analysis, and the detection of manipulative or fraudulent activities are now essential tools for protecting market integrity. However, balancing the need for effective surveillance with ethical considerations like privacy and free speech remains a key challenge. As technology evolves, so too must the tools and strategies used to monitor and safeguard financial markets in the age of social media.

4.13 NLP for Regulatory Compliance

With the increasing complexity of financial markets due to globalization, technological innovations, and the proliferation of social media, the need for advanced and adaptive surveillance techniques is more important than ever. By employing AI-driven systems, cross-market cooperation, and real-time monitoring of social media, regulators can continue to safeguard the integrity of financial markets and protect investors from fraud and manipulation.

The sources provide several compelling examples of how NLP is being applied to enhance various aspects of financial compliance:

- **Regulatory Compliance**: The ever-growing volume and complexity of regulatory requirements pose a significant challenge for financial institutions. AI solutions, such as IBM's Watson Compare & Comply, can leverage natural language processing to map regulatory rules to institutional data and automate the monitoring of compliance obligations. This can significantly reduce the burden of manual compliance processes and enhance the accuracy and efficiency of regulatory reporting.
- **Fraud Detection**: Fraud poses a significant threat to financial institutions and their customers. AI-powered fraud detection systems can analyze transaction data in real-time, identifying anomalies and patterns that may suggest fraudulent activity. Machine learning algorithms can learn from historical fraud patterns and adapt to new fraud techniques, improving the accuracy and efficiency of fraud prevention measures.
- **Market Risk Management**: AI can enhance market risk management by providing more accurate and dynamic assessments of market risks. Machine learning algorithms can analyze real-time market data, including price movements, trading volumes, and news sentiment, to predict market volatility and potential downturns. This allows financial institutions to adjust their investment strategies

and risk mitigation measures proactively, reducing their exposure to market fluctuations.
- **Credit Risk Management**: Assessing credit risk is crucial for lenders to make informed decisions about loan approvals. AI-driven credit scoring systems can analyze a wider range of data points, including alternative data sources like social media activity and online behavior, to provide more comprehensive credit risk assessments. This can help lenders make more accurate predictions about a borrower's likelihood of default and reduce the risk of loan losses.

4.13.1 The Future of AI in Financial Compliance

The adoption of NLP in financial compliance is still in its early stages, but it is rapidly evolving and poised to play an even more significant role in the future. As NLP technologies mature and become more sophisticated, we can expect to see even more innovative applications in areas such as:

- **Predictive Compliance**: AI can be used to predict potential compliance issues before they occur, allowing financial institutions to take proactive measures to mitigate risks and avoid costly penalties.
- **Personalized Compliance**: AI can be used to tailor compliance processes to the specific risks and characteristics of individual customers, reducing the burden of compliance while enhancing its effectiveness.
- **Regulatory Technology (RegTech)**: RegTech refers to the use of technology to enhance regulatory compliance processes. AI is a key enabler of RegTech solutions, providing capabilities such as automated reporting, real-time monitoring, and risk prediction.

While NLP offers immense potential for transforming financial compliance, it's essential to acknowledge that NLP is not a silver bullet. Financial institutions must address the challenges associated with adoption of NLP and use it responsibly and ethically. Collaboration between regulators, financial institutions, and NLP experts is crucial to ensure that it is used to foster a more robust, transparent, and equitable financial system.

4.14 Future Directions

The research points towards a future where the NLP plays an increasingly crucial role in shaping economic forecasting. The pursuit of more robust, accurate, and context-aware sentiment analysis models will continue to drive innovation and improve the effectiveness of this valuable tool in the financial domain. Future research directions in this field are likely to focus on several key areas.

- **Developing more sophisticated algorithms:** Advancements in NLP, particularly in areas like deep learning and contextual language models, will lead to more sophisticated algorithms capable of handling nuanced language, including sarcasm, irony, and domain-specific terminology.
- **Incorporating real-time sentiment data:** Integrating real-time sentiment data from sources like social media and news feeds into trading platforms could provide investors with more dynamic and responsive insights into market sentiment, potentially enhancing their decision-making capabilities.
- **Multi-modal analysis:** Exploring multi-modal analysis, which combines text, images, and other data sources, could provide a more comprehensive understanding of market sentiment. For instance, analyzing images and videos alongside textual data could reveal subtle emotional cues that might be missed by text-only analysis.
- **Explainable AI:** As NLP models become more complex, the need for explainable AI becomes increasingly important. Understanding how these models arrive at their predictions, particularly in the context of financial decision-making, is crucial for building trust and ensuring responsible use.
- **Inclusion of Diverse Instruments:** Financial markets are characterized by a wide range of instruments, including stocks, bonds, derivatives, and cryptocurrencies. Models should incorporate these diverse instruments to develop comprehensive surveillance systems that can effectively monitor all types of transactions and regulate compliances.
- **Multi-lingual analysis:** The complexity of multilingual data processing in financial market surveillance stems from the need to handle diverse languages with varying levels of support in natural language processing tools. Many surveillance systems struggle with less common languages or dialects, leading to potential gaps in monitoring. Models that could analyse data mentioned in any language should be developed.

4.15 Conclusion

Financial market monitoring is a critical task to ensure the integrity and fairness of financial markets. However, it faces significant challenges due to the complexity of financial instruments, technological advancements, globalized markets, and data volume. Financial sentiment analysis can be a valuable tool to complement traditional surveillance methods by analyzing textual data to gauge market sentiment and identify potential risks. By developing automated surveillance systems using NLP, regulators can enhance their ability to detect and prevent misconduct in financial markets, ultimately protecting investors and maintaining market stability. Social media has become a powerful force in the financial markets, influencing stock prices, investor sentiment, and market trends. As its role continues to grow, regulators and financial institutions must adapt by integrating social media surveillance into their broader market oversight frameworks. Real-time monitoring of social media discussions, sentiment analysis, and the detection of manipulative or fraudulent activities are now essential tools for protecting market integrity.

In conclusion, the ongoing research paints a compelling picture of how NLP is transforming the field of economic forecasting. While challenges remain, the ability to extract and analyze sentiment from textual data offers a powerful new tool for understanding and predicting economic trends. As research in this area continues to advance, we can expect to see even more innovative applications of NLP, leading to more informed and insightful economic predictions that benefit investors, policymakers, and businesses alike.

References

1. Srijiranon, K., Lertratanakham, Y., & Tanantong, T. (2022). A hybrid framework using PCA, EMD and LSTM methods for stock market price prediction with sentiment analysis. Applied Sciences, 12(21), 10823.
2. Li, H., Peng, Q., Wang, X., Mou, X., & Wang, Y. (2023). SEHF: A Summary-Enhanced Hierarchical Framework for Financial Report Sentiment Analysis. IEEE Transactions on Computational Social Systems.
3. Long, W., Gao, J., Bai, K., & Lu, Z. (2024). A hybrid model for stock price prediction based on multi-view heterogeneous data. Financial Innovation, 10(1), 48.
4. Liu, W. J., Ge, Y. B., & Gu, Y. C. (2024). News-driven stock market index prediction based on trellis network and sentiment attention mechanism. Expert Systems with Applications, 250, 123966.
5. Zhang, Q., Zhang, Y., Bao, F., Liu, Y., Zhang, C., & Liu, P. (2024). Incorporating stock prices and text for stock movement prediction based on information fusion. Engineering Applications of Artificial Intelligence, 127, 107377.
6. Corizzo, R., & Rosen, J. (2024). Stock market prediction with time series data and news headlines: a stacking ensemble approach. Journal of Intelligent Information Systems, 62(1), 27-56.
7. Du, K., Xing, F., Mao, R., & Cambria, E. (2023, December). FinSenticNet: A concept-level lexicon for financial sentiment analysis. In 2023 IEEE Symposium Series on Computational Intelligence (SSCI) (pp. 109-114). IEEE.
8. Tan, K. L., Lee, C. P., & Lim, K. M. (2023). A survey of sentiment analysis: Approaches, datasets, and future research. *Applied Sciences*, *13*(7), 4550.
9. Yang, S., Ding, Y., Xie, B., Guo, Y., Bai, X., Qian, J., ... & Ren, J. (2023). Advancing Financial Forecasts: A Deep Dive into Memory Attention and Long-Distance Loss in Stock Price Predictions. *Applied Sciences*, *13*(22), 12160.
10. Memiş, E., Akarkamçı, H., Yeniad, M., Rahebi, J., & Lopez-Guede, J. M. (2024). Comparative study for sentiment analysis of financial tweets with deep learning methods. *Applied Sciences*, *14*(2), 588.
11. Frohmann, M., Karner, M., Khudoyan, S., Wagner, R., & Schedl, M. (2023). Predicting the price of Bitcoin using sentiment-enriched time series forecasting. *Big Data and Cognitive Computing*, *7*(3), 137.
12. Smatov, N., Kalashnikov, R., & Kartbayev, A. (2024). Development of Context-based Sentiment Classification for Intelligent Stock Market Prediction. *Big Data and Cognitive Computing*, *8*(6), 51.
13. Roumeliotis, K. I., Tselikas, N. D., & Nasiopoulos, D. K. (2024). LLMs and NLP Models in Cryptocurrency Sentiment Analysis: A Comparative Classification Study. *Big Data and Cognitive Computing*, *8*(6), 63.

14. Kumar, A., Sangwan, S. R., Singh, A. K., & Wadhwa, G. (2023). Hybrid deep learning model for sarcasm detection in Indian indigenous language using word-emoji embeddings. *ACM Transactions on Asian and Low-Resource Language Information Processing*, *22*(5), 1-20.
15. Kumar, A., & Garg, G. (2019). Sentiment analysis of multimodal twitter data. *Multimedia Tools and Applications*, *78*, 24103-24119.
16. Albahli, S., Irtaza, A., Nazir, T., Mehmood, A., Alkhalifah, A., & Albattah, W. (2022). A machine learning method for prediction of stock market using real-time twitter data. *Electronics*, *11*(20), 3414.
17. Kumar, A., & Jaiswal, A. (2019). Swarm intelligence based optimal feature selection for enhanced predictive sentiment accuracy on twitter. *Multimedia Tools and Applications*, *78*, 29529-29553.
18. Kumar, A., & Jaiswal, A. (2020). Systematic literature review of sentiment analysis on Twitter using soft computing techniques. *Concurrency and Computation: Practice and Experience*, *32*(1), e5107.
19. Chen, W., Rabhi, F., Liao, W., & Al-Qudah, I. (2023). Leveraging state-of-the-art topic modeling for news impact analysis on financial markets: a comparative study. *Electronics*, *12*(12), 2605.
20. Kumar, A., & Rani, R. (2016, October). Sentiment analysis using neural network. In *2016 2nd International Conference on Next Generation Computing Technologies (NGCT)* (pp. 262-267). IEEE.
21. Jain, R., & Vanzara, R. (2023). Emerging Trends in AI-Based Stock Market Prediction: A Comprehensive and Systematic Review. *Engineering Proceedings*, *56*(1), 254.
22. Kumar, A., Sharma, K., & Sharma, A. (2021). Genetically optimized Fuzzy C-means data clustering of IoMT-based biomarkers for fast affective state recognition in intelligent edge analytics. *Applied Soft Computing*, *109*, 107525.
23. Han, M., & Zhou, J. (2022). Multi-scale characteristics of investor sentiment transmission based on wavelet, transfer entropy and network analysis. *Entropy*, *24*(12), 1786.
24. Kumar, A., Bhatia, M. P. S., & Sangwan, S. R. (2022). Rumour detection using deep learning and filter-wrapper feature selection in benchmark twitter dataset. *Multimedia Tools and Applications*, *81*(24), 34615-34632.
25. Liapis, C. M., Karanikola, A., & Kotsiantis, S. (2023). Investigating deep stock market forecasting with sentiment analysis. *Entropy*, *25*(2), 219.
26. Leung, M. F., Chan, L., Hung, W. C., Tsoi, S. F., Lam, C. H., & Cheng, Y. H. (2023). An intelligent system for trading signal of cryptocurrency based on market tweets sentiments. *FinTech*, *2*(1), 153-169.
27. Kumar, A. K. S. H. I., & Sharma, A. D. I. T. I. (2019). Systematic literature review of fuzzy logic based text summarization. *Iranian journal of fuzzy systems*, *16*(5), 45-59.
28. Kumar, A., Sharma, A., Sharma, S., & Kashyap, S. (2017, July). Performance analysis of keyword extraction algorithms assessing extractive text summarization. In *2017 International Conference on Computer, Communications and Electronics (Comptelix)* (pp. 408-414). IEEE.
29. Prasad, S., Mohapatra, S., Rahman, M. R., & Puniyani, A. (2022). Investor sentiment index: a systematic review. *International Journal of Financial Studies*, *11*(1), 6.
30. Kumar, A., Sharma, A., & Nayyar, A. (2020, February). Fuzzy logic based hybrid model for automatic extractive text summarization. In *Proceedings of the 2020 5th International Conference on Intelligent Information Technology* (pp. 7-15).
31. Kumar, A., Sharma, A., & Arora, A. (2019). Anxious depression prediction in real-time social data. *arXiv preprint arXiv:1903.10222*.
32. Farimani, S. A., Jahan, M. V., & Milani Fard, A. (2022). From text representation to financial market prediction: A literature review. *Information*, *13*(10), 466.
33. Liapis, C. M., & Kotsiantis, S. (2023). Temporal Convolutional Networks and BERT-Based Multi-Label Emotion Analysis for Financial Forecasting. *Information*, *14*(11), 596.
34. Ranjan, R., & Sharma, A. (2019). Evaluation of frequent itemset mining platforms using apriori and fp-growth algorithm. *arXiv preprint arXiv:1902.10999*.

35. McCarthy, S., & Alaghband, G. (2023). Enhancing financial market analysis and prediction with emotion corpora and news co-occurrence network. *Journal of Risk and Financial Management, 16*(4), 226.
36. Wu, S., & Gu, F. (2023). Lightweight Scheme to Capture Stock Market Sentiment on Social Media Using Sparse Attention Mechanism: A Case Study on Twitter. *Journal of Risk and Financial Management, 16*(10), 440.
37. Cristescu, M. P., Nerisanu, R. A., Mara, D. A., & Oprea, S. V. (2022). Using market news sentiment analysis for stock market prediction. *Mathematics, 10*(22), 4255.
38. Di Tollo, G., Andria, J., & Filograsso, G. (2023). The predictive power of social media sentiment: Evidence from cryptocurrencies and stock markets using nlp and stochastic anns. *Mathematics, 11*(16), 3441.
39. Sharma, A., Sharma, K., & Kumar, A. (2023). Real-time emotional health detection using fine-tuned transfer networks with multimodal fusion. *Neural computing and applications, 35*(31), 22935-22948.
40. Cristescu, M. P., Mara, D. A., Nerişanu, R. A., Culda, L. C., & Maniu, I. (2023). Analyzing the Impact of Financial News Sentiments on Stock Prices—A Wavelet Correlation. *Mathematics, 11*(23), 4830.
41. Fu, K., & Zhang, Y. (2024). Incorporating Multi-Source Market Sentiment and Price Data for Stock Price Prediction. *Mathematics, 12*(10), 1572.
42. Velu, S. R., Ravi, V., & Tabianan, K. (2023). Multi-lexicon classification and valence-based sentiment analysis as features for deep neural stock price prediction. *Sci, 5*(1), 8.
43. Ranjan, R., & Sharma, A. (2020). Voice-controlled IoT devices framework for smart home. In *Proceedings of First International Conference on Computing, Communications, and Cyber-Security (IC4S 2019)* (pp. 57-67). Springer Singapore.
44. Ouyang, Z., & Lu, M. (2024). Systemic Financial Risk Forecasting with Decomposition–Clustering-Ensemble Learning Approach: Evidence from China. *Symmetry, 16*(4), 480.
45. Dogra, V., Alharithi, F. S., Álvarez, R. M., Singh, A., & Qahtani, A. M. (2022). NLP-Based application for analyzing private and public banks stocks reaction to news events in the Indian stock exchange. *Systems, 10*(6), 233.

Chapter 5
Managing Uncertainty in NLP: Advanced Techniques and Approaches

Ravleen Kaur and M. P. S. Bhatia

Abstract Uncertainty is an inherent challenge in Natural Language Processing (NLP), affecting various stages of model development and application, from data collection to prediction and interpretation. This chapter explores the concept of uncertainty within the context of NLP, examining its impact on language models, their training, and their real-world deployment. It addresses the different types of uncertainty—epistemic and aleatoric—and their manifestation in NLP tasks such as text classification, named entity recognition, and machine translation. The chapter also highlights techniques for quantifying and mitigating uncertainty, including Bayesian methods, Calibration methods, and Conformal prediction. Furthermore, we discuss the implications of uncertainty for model interpretability, robustness, and fairness, offering insights into how incorporating uncertainty can lead to more reliable and transparent NLP systems. Through case studies and practical examples, this chapter provides a comprehensive understanding of how uncertainty affects NLP and outlines strategies for managing it in real-world applications.

Recent advancements in Natural Language Processing (NLP) technologies have significantly changed how machines understand and work with language, allowing them to interpret and generate text with impressive accuracy. NLP emerged from simple text processing in its early days to the current dominance by deep learning and large-scale models. Because of these advancements, several applications that add value to communication and business operations and recover insights from unstructured data have developed. NLP thus has been a rapidly expanding beneficiary in health care, finance, customer service, and social media analysis, which thereby changed, at least, business operations and day-to-day life. NLP aims to bridge the gap between human language and machine understanding, but one of its fundamental challenges is managing the uncertainty that pervades human communication. Uncertainty arises from the ambiguity, variability, and probabilistic nature of language, making it challenging for models to generate or comprehend text with consistent accuracy [1, 2]. Natural language is intrinsically complex because it

R. Kaur (✉) · M. P. S. Bhatia
Netaji Subhas University of Technology, New Delhi, Delhi, India
e-mail: ravleen.kaur.phd22@nsut.ac.in; mps.bhatia@nsut.ac.in

A. Kumar, S. R. Sangwan (eds.), *Transformative Natural Language Processing*,
https://doi.org/10.1007/978-3-031-88988-2_5

remains highly ambiguous, variable, and context dependent. Hence, it becomes prone to misinterpretation and inaccuracy concerning understanding the intent of the users. For example, words may have several senses, phrases are syntactically ambiguous, and the meaning of a sentence often depends on contextual clues which are not explicitly mentioned at all. Furthermore, since language is constantly alive and with the ever-changing aspects of cultural phenomena, idioms, technical terms, etc., add another degree of complexity in designing practical NLP systems. This ambivalence can lead to very serious application repercussions, for example, in sentiment analysis, where reading an emotion wrong may result in misguided responses, or even in legal text processing wherein accuracy is paramount. This chapter provides a comprehensive introduction to the idea of uncertainty in NLP, including its sources, implications for different NLP applications, and methods for addressing it. The chapter concludes with a summary of the organization of the book and a roadmap for techniques and strategies for mastering uncertainty management in NLP.

5.1 Defining Uncertainty

Language is a rich, complex system characterized by context-dependent meanings, subtle cultural variations, and potential for multiple interpretations. Furthermore, language is continually evolving, with new terms and usages emerging over time. As a result, NLP systems must contend with various forms of uncertainty when processing natural language data. The data encountered by NLP models often includes noise, such as typos, informal language, or domain-specific jargon, further complicated interpretation [3]. Several key sources of uncertainty consist of [4]:

- **Ambiguity**: Words and phrases can have multiple interpretations. The meaning of words, phrases, and even entire sentences can vary according to the context in which they appear. For example, the term "*bank*" can refer to a financial institution or the side of a river; the appropriate interpretation depends on the context. Similarly, the phrase "*He saw her duck*" could either mean he saw her pet bird or her lowering her head to avoid something. Effectively navigating ambiguity is crucial for NLP applications like machine translation, sentiment analysis, and conversational agents, where understanding precise meanings is essential for delivering accurate results.
- **Data noise**: Errors, inconsistencies, and irrelevant information are common in textual data. This is especially true for content created by users, such as social media posts, where grammatical irregularities, abbreviations, slangs, and spelling mistakes are frequently used. For example, people may type "*u*" instead of "*you*", substitute words with emojis, or utilize expressions that are widely used in some areas but may not be understood in other communities. Additionally, varying degrees of language proficiency among users can introduce inconsistencies in sentence structure and vocabulary, further complicating the data. Unrelated

data might cause NLP algorithms to get confused since it adds an additional layer of noise in the form of off-topic remarks or advertising materials.

- **Variability across domains**: A model's performance may be impacted if it encounters language patterns or terminologies that it was not trained on because language use varies greatly between various domains (e.g., scientific writing vs. social media). For example, a model trained primarily on news articles would not be able to comprehend social media messages that contain abbreviations or slang, such as "*lol*" (laugh out loud) or "*brb*" (be right back). On the other hand, a model that was trained on technical documents can misinterpret informal conversations or fail to pick up on nuances of sarcasm or humor that are common in normal speech. This domain variability can lead to inaccuracies in tasks such as sentiment analysis, topic classification, or language translation.
- **Out-of-Distribution Data**: A model's ability to predict outcomes accurately may be affected when it is exposed to data that is not consistent with its training set, hence increasing uncertainty. For example, if an NLP model was trained primarily on formal texts like news articles and academic papers, it may encounter difficulties when presented with informal content from social media, which often includes slang, abbreviations, and emotional expressions. Similarly, a model trained in specific domains, such as medical literature, might falter when faced with consumer reviews or casual conversations that use different terminologies and linguistic styles. The mismatch between the training data and the new data can lead to a lack of confidence in the model's predictions, resulting in errors or misinterpretations.

5.1.1 Why Need Uncertainty Estimates?

In NLP, uncertainty estimations are crucial since mistakes can occur in even the most sophisticated models [5]. Even a machine translation system with 98% accuracy, for instance, occasionally generates inaccurate translations. Estimates of uncertainty are important because they reveal when the results of a model could be less accurate or prone to mistake. By quantifying the confidence level of predictions, these estimates help identify cases that require closer examination or human intervention. This proactive strategy promotes more informed decision-making, improves error management, and ultimately raises the reliability and security of NLP systems for a range of applications. Here is a comprehensive overview of the advantages of these estimates:

- *Building Trust in Predictions:* By giving consumers an idea of the degree of uncertainty in a given forecast, the system is made more trustworthy. For example, in medical diagnosis, a model predicting the likelihood of a disease can indicate a confidence level, allowing doctors to trust the prediction more.
- *Comparing Model Performance:* Model selection is made easier by uncertainty estimations, which allow models to be compared according to dependability

measures and confidence levels. For instance, evaluating models according to their level of uncertainty in sentiment analysis might assist in determining which model is more reliable in ambiguous cases.

- *Identifying Areas of Improvement:* By examining uncertainty estimates, developers may identify particular model components that need to be improved, focusing their efforts on those areas.
- *Listing Reasonable Responses with Probabilistic Guarantees:* Instead of producing a single deterministic result, models may now produce a variety of possible solutions, each linked to its likelihood, thanks to uncertainty estimations. When a question-answering system displays many likely solutions together with their degrees of confidence, users gain a more thorough understanding.
- *Producing Natural Responses in Dialogue Agents:* By allowing dialogue agents to communicate their confidence, uncertainty quantification promotes more relevant and organic interactions. By saying "I'm fairly certain about that," as opposed to "I'm not sure," a virtual assistant can have more human-like discussions.
- *Enhancing User Experience:* Systems may provide users with a more intuitive and knowledgeable experience by customizing replies according to ambiguity, which raises user satisfaction levels overall. Recommendation systems can help users make better decisions by providing them with options along with confidence ratings.
- *Facilitating Active Learning:* By determining which samples are most instructive to query for human annotation, uncertainty estimates can help active learning situations run more smoothly. For example, in image classification, a model may request labels for images it is uncertain about, ensuring effective use of human resources Tables 5.1 illustrates Coca Cola's approach to enhancing its insights into public sentiment. Table 5.2 illustrates how Amazon improved its assessment of customer satisfaction through uncertainty-aware NLP techniques.

5.1.2 Types of Uncertainty in NLP

Comprehending and handling uncertainty is essential in the field of Natural Language Processing (NLP) in order to create efficient models that can correctly generate, interpret, and react to human language. When processing natural language, a variety of uncertainties can arise, each posing a different set of difficulties for NLP systems. Here's a comprehensive explanation of the types of uncertainty in NLP:

- Epistemic Uncertainty (Model Uncertainty)
- Aleatoric Uncertainty (Data Uncertainty)
- Out-of-Distribution Uncertainty
- Label Uncertainty (Fig. 5.1) illustrates the primary types of uncertainty encountered in NLP

Table 5.1 Uncertainty in social media monitoring for sentiment analysis: the case of Coca Cola

Category	Details
Background	Coca-Cola, a leading beverage brand, closely monitors social media to gauge public sentiment around its products, campaigns, and corporate initiatives. The "Taste the Feeling" campaign launched in 2010 targeted younger consumers, with sentiment analysis playing a crucial role in its success
Challenges	1. **Ambiguous Language**: Social media posts often used slang, colloquial terms, or informal language, making it hard to interpret sentiment (e.g., "This new Coke is a vibe!")
	2. **Sarcasm**: Sarcastic remarks (e.g., "Oh great, another sugary drink to ruin my diet!") could mislead sentiment algorithms
	3. **Contextual Changes**: The COVID-19 pandemic created a shifting landscape, with consumer sentiment fluctuating between health concerns and indulgence
	4. **Noise in Data**: Social media content contained irrelevant posts, memes and advertisements that could obscure genuine consumer sentiment
Approach	1. **Data Preprocessing:** Developed a sophisticated data-cleaning pipeline to filter out spam accounts, irrelevant posts, and identify campaign-related content using hashtags
	2. **Contextual Sentiment Analysis Models:** Utilized transformer-based models like BERT to understand the context and sentiment nuances in posts (e.g., distinguishing "This new Coke is fire!" from "That Coke is a disaster!")
	3. **Training with Diverse Data:** Trained models on diverse social media data to handle informal language, slang, and regional expressions
	4. **Incorporating Uncertainty Quantification:** Implemented techniques to assess the confidence level of predictions, flagging low-confidence classifications for manual review
	5. **Real-Time Monitoring:** Established a dashboard to monitor sentiment trends, allowing for quick strategy adjustments based on public reaction
Results	1. **Improved Accuracy**: Reduced misclassifications, particularly for ambiguous or sarcastic tweets, leading to a clearer understanding of campaign sentiment
	2. **Actionable Insights:** Prioritized high-confidence predictions for immediate action and flagged uncertain predictions for further review, helping to mitigate PR risks
	3. **Adaptive Marketing Strategies**: Leveraged real-time sentiment data to quickly pivot marketing strategies. For instance, when health concerns emerged, Coca-Cola adjusted its messaging to emphasize low-sugar products, resonating positively with the audience
Conclusion	Coca-Cola's approach highlights the importance of managing uncertainty in sentiment analysis for social media monitoring. By applying advanced techniques, including contextual modeling and uncertainty estimation, the company enhanced its understanding of public sentiment, leading to more informed decision-making, adaptive marketing strategies, and stronger customer relationships during challenging times

Table 5.2 Handling uncertainty in NLP for customer satisfaction analysis at Amazon

Category	Details	
Background	Amazon, a global e-commerce giant, relies heavily on customer feedback to enhance product quality, improve customer service, and optimize its recommendation system. With millions of customer reviews and support interactions generated daily, analyzing this vast amount of text data accurately is critical for understanding customer satisfaction and improving service	
Challenges	1. Ambiguity in Customer Feedback: Customer reviews often contained ambiguous language or mixed sentiments, such as "The delivery was late, but the product quality was excellent", leading to inaccurate sentiment classification	
	2. Subjectivity in Sentiment: Customers expressed similar sentiments differently, making it difficult to interpret subtle differences (e.g., "Not bad" could be positive or neutral), causing inconsistencies in satisfaction scores	
	3. Sarcasm and Irony Detection: Sarcastic reviews, such as "Oh great, another defective product from Amazon!" were often misclassified by sentiment models, resulting in misleading satisfaction assessments	
	4. Noisy and Unstructured Data: Reviews included spam, irrelevant information, and informal language, adding noise and complicating the analysis process	
Approach	**Preprocessing techniques**	**Data Cleaning:** Filtering spam, irrelevant comments, and advertisements, and normalizing text to handle informal language and errors
		Contextual Analysis: Separating mixed sentiments in single reviews for finer classification
	Contextual Sentiment Models	**Transformer-Based Models (e.g., BERT)**: Fine-tuned for understanding nuanced expressions like sarcasm and mixed feelings
		Aspect-Based Sentiment Analysis (ABSA): Evaluating sentiments toward specific aspects (e.g., delivery vs. product quality)
	Uncertainty Quantification	**Confidence Estimation:** Flagging low-confidence predictions for human review, allowing priority handling
		Ensemble Methods: Aggregating predictions from multiple models to reduce uncertainty and minimize interpretations
	Human-in-the-Loop Review	For high uncertainty reviews, human annotators verified sentiment manually, ensuring accurate labels and continuous model improvement

(continued)

Table 5.2 (continued)

Category	Details
Results	1. Improved Accuracy: The system reduced sentiment classification errors by 30%, particularly in ambiguous or sarcastic cases, leading to more accurate customer satisfaction measurements
	2. Enhanced Customer Service: Uncertainty quantification enabled Amazon to identify problematic cases quickly and resolve customer issues in real-time, improving satisfaction levels
	3. Data-Driven Product insights: Aspect-based sentiment analysis provided feedback on product features, allowing Amazon to optimize specific areas (e.g., packaging improvements based on customer complaints)
	4. Continuous Model Improvement: Human-in-the-loop approach updated training data with accurately labeled examples, leading to ongoing model enhancements in handling uncertainty
Conclusion	Amazon's approach demonstrates the importance of addressing uncertainty in NLP for customer satisfaction analysis. Using advanced techniques like contextual modeling and human-in-the-loop processes, Amazon reduced misclassification and improved customer service

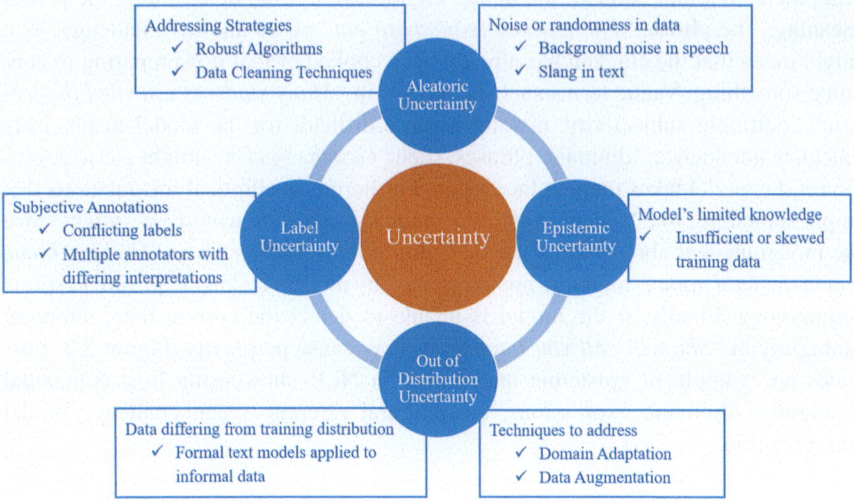

Fig. 5.1 Types of uncertainty

Epistemic Uncertainty (Model Uncertainty)

Model uncertainty, also known as epistemic uncertainty, is essentially about the unknowns around the parameters of the model and its underlying architecture [6]. The main cause of this kind of uncertainty is a lack of knowledge about the connections and patterns the model is supposed to discover. It specifically occurs when the model is trained on a dataset that is too small, too varied, or skewed in a way that makes it difficult for the model to effectively capture the variability seen in real-world data. For example, consider an NLP model intended for sentiment analysis. This model may acquire a distorted concept of sentiment if it is trained on a small dataset that is predominantly composed of positive reviews. This might result in significant uncertainty when the model comes across neutral or negative reviews. The model's capacity to generalize its predictions to fresh, unknown data is hampered by this constraint in the training data, especially when dealing with ambiguous or contextually complex statements.

Epistemic uncertainty can take many different forms. If a statement relies on context that is absent from the training data or uses sophisticated language, the model may have trouble correctly classifying it. For example, in "*He saw the bat fly out of the cave*", the word "*bat*" might refer to a flying animal or a piece of sporting equipment. Without context, the model might have trouble determining the proper meaning. The phrase "*The chicken is ready to eat*" poses another challenge, as it might mean that the chicken has already been cooked or that it is preparing to consume something. Vague terms such as "*many*" in "*Many students attended the lecture*" contribute subjectivity, making it more difficult for the model to precisely calculate attendance. Idiomatic phrases, such "*kick the bucket*," might cause confusion if the model takes them at face value. Furthermore, elliptical formulations that imply something like "*I'd like to have a sandwich and John would too*" might leave the model unclear about John's wants. Cultural references, such as "*He hit a home run in the final game*," might be misinterpreted by models lacking exposure to sports terminology. Finally, if the model is unable to detect the current date, temporal ambiguity in "*She will call you tomorrow*" may cause perplexity. Figure 5.2 illustrates an example of epistemic uncertainty in NLP, showcasing how contextual ambiguity, idiomatic expressions, and cultural references can challenge model interpretation.

Aleatoric Uncertainty (Data Uncertainty)

Data uncertainty, or aleatoric uncertainty, refers to the intrinsic noise and randomness in the data that cannot be completely removed, not even with the best modeling strategies. This kind of uncertainty, which results from unpredictable variations that have the potential to skew data and impair model performance, is a major problem in NLP. Aleatoric uncertainty, for instance, might happen in the context of speech recognition when a system misinterprets spoken words because of background noise or overlapping speech from several speakers. These external variables might result in inaccurate transcriptions, which makes it challenging for the model to

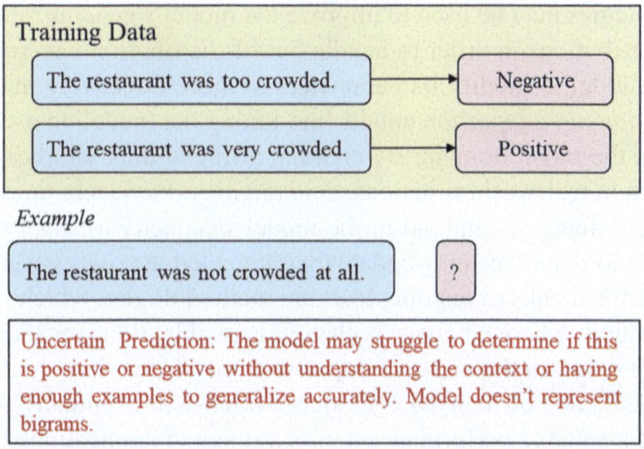

Fig. 5.2 BOW for sentiment analysis

correctly understand the intended message. Similarly, misspellings and improper punctuation in written text add noise that can confuse models that are taught to interpret language. Social media posts are a prime example of aleatoric ambiguity since they are full of slang, emoticons, acronyms, and casual language, all of which can have quite different meanings depending on the context and usage. For NLP systems, navigating the subtleties of human communication while managing the inherent variability in the data becomes more difficult due to this unpredictability. Aleatoric uncertainty must thus be addressed throughout the data collecting and preparation phases. This includes employing noise-robust algorithms, implementing strategies that can adjust to data unpredictability, and utilizing robust data-cleaning approaches. Through recognition and consideration of aleatoric uncertainty, practitioners may improve the robustness of their natural language processing (NLP) models, resulting in more dependable outcomes in practical scenarios.

Out-of-Distribution Uncertainty

When a model comes into data that substantially differs from the distribution it was trained on, it is said to be experiencing out-of-distribution uncertainty, which makes its predictions less reliable. When NLP models are used in unfamiliar situations or domains for which they have not been sufficiently trained, this kind of ambiguity frequently develops. An NLP model trained mostly on formal written text, for example, can find it difficult to understand social media postings and text messages effectively due to their informal style. The informal interactions contain stylistic variations, slang, acronyms, and other grammatical structures that might confuse the model and raise uncertainty in its results. Consequently, this might lead to the model misclassifying feelings, misinterpreting requests, or failing to grasp contextual subtleties, all of which could have serious ramifications for applications such as content moderation, sentiment analysis, and customer assistance.

Several strategies may be used to improve the model's generalization across different data distributions in order to handle out-of-distribution uncertainty. In order to enable the model to modify its parameters to more effectively manage the new data features, *domain adaptation* entails fine-tuning the model on a smaller dataset that represents the target domain. By creating artificial data samples that replicate the variety seen in real-world situations, *data augmentation* techniques can increase the size of the training set and aid in the model's capacity to adapt to unexpected inputs. Furthermore, models may update their knowledge progressively as new data becomes available thanks to ongoing learning methodologies, which is very helpful in dynamic contexts where language is changing quickly. By using these techniques, practitioners may lessen the negative impacts of out-of-distribution uncertainty and increase the resilience of their NLP models, which will eventually result in more accurate and dependable performance across a range of applications.

Label Uncertainty

Tasks involving subjective assessments, such sentiment analysis, text categorization, and other annotation procedures, give rise to label ambiguity. This kind of ambiguity arises when several annotators, each reflecting their own biases and perceptions, give conflicting labels for the same data. In a sentiment analysis job, for example, one annotator could identify a product review as "positive" because they emphasize the exuberant tone, whereas another annotator would identify the same review as "neutral" since it contains a range of attitudes. This disparity draws attention to how difficult it may be to establish the ground truth, which makes machine learning model training difficult.

Conflicting labels during training can make it difficult for the models to pick up reliable representations of the underlying data, which lowers performance and increases prediction uncertainty. The model may come across examples with different interpretations and be unable to settle on a trustworthy view of the data as a result of inconsistent labeling, which might cause confusion. Many strategies may be used to reduce label ambiguity. Using several annotators for each input and aiming for agreement on the final label is an efficient method that can contribute to the creation of a more reliable ground truth. As an alternative, probabilistic labeling captures the inherent ambiguity in subjective judgments by representing label uncertainty as a probability distribution across potential labels rather than a single label.

5.1.3 The Importance of Managing Uncertainty in NLP

Managing uncertainty in natural language processing is not simply a technical challenge but is important for the credibility, understandability, and effective functioning of NLP systems in practice. From the users' perspective, any NLP system is expected to perform adequately and produce the right information at the right place. The failure to manage uncertainty can undermine user trust and inhibit further use. For instance, consider the case in healthcare environments where an NLP application system misinterprets handwritten clinical notes leading to inaccurate diagnosis or poor clinical interventions which affect patient care. Similar issues can be observed in some financial applications where there is a possibility of predicting changing market trends using sentiments; an incorrect analysis of the global investors' moods may lead to purchase or dumps at the wrong time thus leading to losses.

In order to effectively respond to these issues, the area of NLP has already begun to incorporate more sophisticated techniques and strategies that improve the accuracy and reliability of the models while addressing the complexities inherent in natural language. This chapter is concerned with the two primary strategies: The integration of knowledge-based systems and the utilization of innovative machine learning models.

5.2 Knowledge-Based Systems to Improve NLP Accuracy

Natural language texts are an ubiquitous input to a spectrum of information processing applications. Most of these applications not only need analysis on the surface structure but also require extraction and understanding about meanings behind it. For example, automatic document classification or categorization in databases and search engines significantly profit from understanding the semantic content of text. In order for a natural language processing (NLP) system to be able to make sense and work with meanings behind text, it requires some understanding of the real world as well domain from discourse. The knowledge-based approach to NLP focuses on the methods for acquiring and representing such knowledge, as well as applying it to address well-established challenges in NLP, such as ambiguity resolution. In this chapter, we will examine knowledge-based solutions for various NLP problems, explore a range of applications for these solutions, and highlight several exemplary systems developed within the knowledge-based paradigm.

A piece of natural language text can be interpreted as a set of indicators that convey its meaning, with these indicators structured according to the rules and conventions of the language and the stylistic choice of its authors. These indicators include elements like words, their inflections, the sequence in which they appear, punctuation and more. It is well understood that these indicators typically do not provide a direct link to a single, definitive meaning. Instead, they can suggest multiple possible interpretations for a given text, accompanied by various ambiguities and gaps. To discern the most appropriate meaning from an input text, the NLP system must utilize several types of knowledge, including an understanding of the natural language itself, the specific domain of discourse, and broader world knowledge.

5.2.1 Domain-Specific Knowledge Integration in NLP

The majority of knowledge-based NLP systems are made to function in specific domains. To process texts relevant to that topic, these systems make use of in-depth domain-specific expertise. Domain-specific systems typically process natural language more easily for several reasons:

- **Reduction of Ambiguities:** By choosing a specialized domain, which prioritizes domain-specific interpretations over general or colloquial ones, many of the ambiguities present in natural language are lessened. This simplifies the process of understanding and gets rid of conflicting interpretations.
- **Narrowed Inferential Scope:** By reducing the number of potential inferences, the knowledge of the selected domain helps to fill knowledge gaps. Systems can employ more focused reasoning strategies that are specifically pertinent to the topic at hand thanks to this emphasis.
- **Feasibility of Knowledge Encoding:** Attempting to capture the large amount of knowledge that exists worldwide is significantly more difficult than encoding specific knowledge related to a limited field. Therefore, by including expert information—which might take the form of rules, heuristics, or case-based knowledge based on past experiences—domain-specific systems can provide

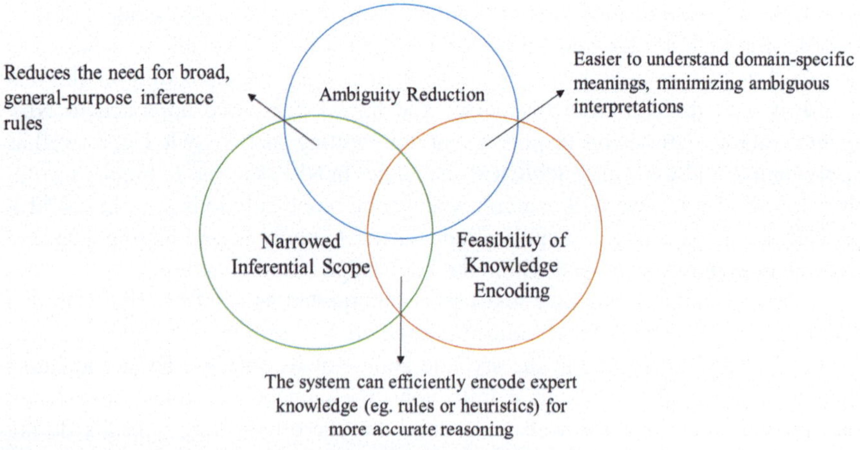

Fig. 5.3 Domain specific knowledge integration

richer and more efficient models of understanding. Figure 5.3 illustrates the integration of domain-specific knowledge in NLP systems, demonstrating how ontologies and knowledge graphs enhance contextual understanding.

Domain Specific Knowledge Graphs and Ontologies

The construction of domain-specific knowledge graphs from specialized corpora is essential for tackling issues within certain domains, even though generic and open-world knowledge graphs have been widely used to solve a variety of domain-independent activities. This requirement results from domain-specific knowledge graphs' inherent connections to key applications and issues within that area. According to some research, domain-specific knowledge graphs are a unique kind of knowledge graph designed to capture the complexities of certain domains. According to some, these knowledge graphs are the result of improving an already-existing domain ontology, which offers a fundamental framework for expressing knowledge in that domain. Knowledge graphs are graph-based data structures that represent entities (nodes), and their interrelationships (edges), in a manner that reflects real-world connections. A knowledge graph usually has nodes that correspond to entities, which might be things, thoughts, events, or even more abstract notions. The interactions between these entities are represented by the edges, and they can be classified as causal (e.g., "causes"), associative (e.g., "related to"), or hierarchical (e.g., "is a").

As semantic data models, ontologies provide the types of things that exist in each domain as well as the attributes that are used to characterize them. Instead of describing examples or people within the field, they are generic frameworks intended to capture large categories of concepts that have similar traits. For example, ontology concentrates on the generic class of "dogs," capturing the traits that apply to many dogs, rather than describing the particular characteristics of a single dog called Spot. This method makes it easier to reuse the ontology in the future to represent more dog instances.

Although domain specificity has its benefits, certain applications do not allow for such strict limitations. Texts from a certain domain may include a wide variety of topics outside of that domain, or a system may need to be able to take any text as input. Giving the system expert knowledge in every possible field becomes challenging in these situations. Rather, a foundation of commonsense knowledge that includes broad world ideas must be established. Addressing problems like ambiguity and variation in text interpretation can be greatly aided by this commonsense understanding. Knowledge-based solutions have had a significant impact on NLP, providing a competitive alternative to methods that just use grammar and linguistic information. Knowledge-based systems make it easier to integrate NLP with other knowledge-based AI systems, such those that concentrate on logical reasoning, engineering design, and problem-solving. The focus is shifted from only natural language front ends to more integrated systems where NLP and other AI activities work together to improve each other's capabilities. Thus, Knowledge Based NLP systems allow AI solutions to be based on natural language representations of real-world inputs and outputs. Additionally, a wide range of applications requiring the processing of various languages have been emphasized by Knowledge Based NLP systems. For activities requiring multilingual processing, such as machine translation, multilingual database querying, information retrieval, and multilingual summarization, knowledge-based techniques are especially beneficial. Systems that handle more than two languages greatly benefit from a common, interlingual representation of meaning since linguistic structures and norms vary greatly among languages. Such language-independent meaning representations are derived and manipulated in a way that is very compatible with the knowledge-based approach to natural language processing.

Rule Based Systems and Expert Databases

Rule-Based Systems work based on clearly stated rules that control how language inputs are processed. Usually written as "if-then" statements, these rules specify how the system will react to certain language situations. The use of rule-based systems is especially beneficial in situations when accuracy is crucial, like:

- *Parsing Structured Text*: Rule-based systems may correctly extract information from well-defined, structured formats (like XML or JSON) by applying syntactic and semantic rules. These rules' deterministic character reduces ambiguity and improves parser consistency.
- *Information Extraction:* Predefined criteria can be used to recognize and categorize entities or particular information from a text in tasks like named entity recognition (NER) or event extraction. A regulation may state, for instance, that any capitalized term that comes after "Inc." qualifies as an organization.

However, while rule-based systems offer control and clarity, they frequently have trouble with the ambiguities, idioms, and contextual variances that are a part of real language. This limitation arises due to the need to cover many linguistic variations, which can be cumbersome and could result in an overabundance of rules.

Text analysis may make use of expert databases, which are specialized repositories that hold domain-specific information. These databases are made up of carefully chosen facts, statistics, and heuristics that provide background knowledge pertinent to specific fields. Expert databases in NLP can improve the way ambiguous language inputs are processed by:

- *Providing Contextual Knowledge*: The system may retrieve domain-relevant knowledge by querying the expert database while text analysis is underway. To resolve ambiguities in patient records, for example, a medical NLP application can make use of an expert database that includes medication interactions, symptoms, and therapy recommendations.
- *Facilitating Knowledge Enrichment:* Systems can lessen uncertainty by combining NLP processing with expert knowledge. To further contextualize the analysis and enhance comprehension, the system may, for instance, access the expert database to determine whether a prescription mentioned in a phrase is an antibiotic.

5.2.2 Enhancing NLP with Expert Systems

Expert systems are a significant development in artificial intelligence that are intended to mimic human experts' decision-making skills in particular domains. These systems can solve complicated issues and make well-informed judgments in a variety of fields, like healthcare, economics, law, and engineering, by methodically using a structured framework of rules, factual knowledge, and advanced inference techniques. Expert systems' fundamental architecture usually consists of two primary parts:

- *Knowledge Base:* An extensive collection of domain-specific information, such as factual information, heuristics, and regulations governing the connections between various ideas within the domain, is referred to as a knowledge base. The expert system is built on the knowledge base, which allows it to reason about different situations and draw conclusions from known data.
- *Inference Engine:* This critical component uses inference rules to apply logical reasoning to the knowledge base in order to make judgments and draw inferences from the input. The inference engine can use a number of reasoning strategies, such as backward chaining, which goes backward from a goal to locate supporting evidence, and forward chaining, which draws conclusions from known facts.

Inference Engines for Uncertainty Management

The inference engine, a crucial part of expert systems, is in charge of drawing inferences from the data and guidelines stored in the knowledge base. By using sophisticated reasoning strategies like fuzzy logic and probabilistic reasoning, these engines can efficiently handle uncertainty in NLP applications.

Calibration: A Frequentist Perspective

Calibration in machine learning refers to the degree to which the predicted probabilities of a model reflect the true likelihood of correctness. In a well-calibrated model, if we denote the confidence level as the predicted probability of a prediction being correct, then the model is considered calibrated if the following condition holds:

$$P(model\ is\ correct \mid confidence\ is\, \alpha = \alpha$$

For any given confidence level α (where $0 \le \alpha \le 1$), the proportion of predictions that are correct should approximately equal α. This means that if the model predicts that a particular prediction is correct with a confidence of α, then we expect that approximately $\alpha \times 100\%$ of all such predictions should indeed be correct.

Calibration: Under and Over Confidence

Calibrated Confidence: A model is considered calibrated when the predicted probabilities of its predictions accurately reflect the true likelihood of correctness. For example, if a model assigns a confidence of 0.80 to its predictions, we expect that about 80% of those predictions to be correct. Calibration ensures that the model's confidence levels are trustworthy and the users can rely on the probabilities it provides for decision-making. A calibrated model effectively balances its confidence, providing accurate estimates without being overly cautious or overly bold.

Under-confidence: When the model's predicted probabilities are lower than the actual likelihood of correctness. For instance, if a model frequently assigns a confidence level of 0.60 to its predictions but is correct 80% of the time, it is underestimating its certainty. This lack of confidence can lead to missed opportunities, as decisions based on these predictions may be overly cautious. Users might fail to act on predictions that are actually more reliable than indicated, potentially leading to negative outcomes in critical situations.

Overconfidence: When a model's predicted probabilities are higher than the actual likelihood of correctness. For example, if a model consistently gives predictions with a confidence of 0.90 but only achieves a 70% accuracy rate, it is misleadingly confident. Overconfidence can result in a false sense of security, leading users to make decisions based on predictions that are less reliable than they appear. This can have severe implications in high-stakes scenarios, such as medical diagnoses or financial investments, where decisions made based on inflated confidence can result in significant risks and consequences.

Estimating Calibration Error

Calibration error measures the discrepancy between a model's predicted confidence levels and its actual accuracy in making predictions. A common technique for estimating this error is through ***binning***, which organizes the predicted probabilities into discrete intervals or "bins". This method enables a more straightforward evaluation of how well a model's confidence *(conf)* corresponds to its accuracy *(acc)*.

$$ECE = \sum_{m=1}^{M} \frac{|B_m|}{n} \left| acc(B_m) - conf(B_m) \right|$$

The measure involves dividing the data into M equally spaced bins, where B represents the bins and m denotes the specific to *M* equally spaced bins, where *B* represents the bins and *m* denotes the specific bin number.

Calibration Method Application

To improve the reliability of predicted probabilities generated by machine learning models, several calibration techniques can be employed. These techniques aim to align the predicted confidence scores with the true probabilities of outcomes, thus enhancing decision making processes. The various techniques are as follows:

Temperature Scaling

Temperature Scaling is a simple yet effective post-hoc calibration technique specifically designed for softmax-based classification tasks. This method adjusts the logits (raw scores) produced by a model to improve the reliability of the predicted probabilities.

Softmax Function: In classification tasks, the softmax function converts the logits from a neural network probability. The basic formulation of the softmax function for a class K is given by:

$$P(y = k|x) = \frac{\exp(S_k / T)}{\Sigma_j \exp(z_j / T)}$$

where z_k are the logits for class *k*, *T* is the temperature parameter.

Temperature Parameter T: It is crucial in modulating the output probabilities:

- Higher values of T: When T is increased (i.e., T > 1), the softmax function produces softer probabilities. This means that the model becomes less confident, distributing probability mass more evenly across classes.
- Lower values of T: When T is decreased (i.e., T < 1), the probabilities become sharpers, leading to more confident predictions where the model assigns higher probabilities to the most likely classes.

Platt Scaling

A calibration technique called platt scaling converts unprocessed prediction scores into calibrated probabilities. This method, which was first put out by Platt in 1999, works especially well for binary classification tasks. It is predicated on the idea that a sigmoidal function may accurately represent the connection between the expected scores and the actual probability.

Sigmoid Mapping: The objective is to model the conditional probability p(y|z) using a sigmoid function:

$$\hat{p}(y = 1|z) \approx \sigma(az + b)$$

where, $\sigma(t) = \frac{1}{1 + e^{-t}}$ is the sigmoid function; *a* and *b* are parameters that need to be trained.

Probabilistic Reasoning

A key concept in natural language processing (NLP) is probabilistic modeling, which is concerned with using probability distributions to describe knowledge and uncertainty [7]. With the use of this technique, NLP systems are able to assess the probability of different results or interpretations given a set of input data, facilitating better decision-making when faced with uncertainty. The main goal of probabilistic modeling in NLP is to represent the inherent uncertainty in language. The meaning of a word or phrase can change depending on the context in which it is used. For instance, think of the term "*bark*", it might mean the barking sound of a dog in one context and the outer bark of a tree in another.

The foundation of probabilistic models is the idea that language may be viewed as a sequence of occurrences with different levels of confidence. A model may, for example, examine a document in a text classification job and generate a probability distribution over several categories, including "sports," "politics," or "entertainment." The model may provide the degree of confidence it has in each group instead of just assigning one, which can be very helpful in situations that are borderline and need complex classification. NLP systems become easier to understand and enable users to make better judgments by providing a quantified sense of confidence levels based on the model's outputs. This section delves into three fundamental ideas: random variables, joint probability distributions, and conditional independence, each of which plays a crucial role in managing uncertainty and ambiguity in linguistic data.

Random Variables

Random variables are crucial components in probabilistic modeling because they capture the inherent uncertainty of language properties or outcomes. There are several possible values for a random variable, and each has a corresponding probability.

Definition and Types

- **Discrete Random Variables:** Variables that can have a countable number of different values are known as discrete random variables. In a text classification task, for instance, a random variable X may indicate whether a certain word appears in a document; it could have the values 0 (absent) or 1 (present).
- **Continuous Random Variables:** Within a specified range, these variables can have an endless number of values. For instance, if a sentence's emotion score were a continuous variable, it may be between −1 (negative) and +1 (positive).

For discrete variables, a Probability Mass Function (PMF) or probability density function (PDF) can be used to define a random variable X. For example, the PMF of X may be described as follows:

$$P(X = x) = p_x$$

where p_x is the probability of the variable X taking the value x.

Joint Probability Distribution

A Conditional Probability Distribution (CPD), which measures the connection between a node and its parent nodes in the graph, is linked to each node in the network. A joint probability distribution P(X) over a collection of variables $X = \{X_1, X_2, \ldots, X_n\}$ is represented formally by a Bayesian network, where the joint distribution may be factorized into a product of Conditional Probabilities, given by:

$$P(X) = \prod_{i=1}^{n} P(X_i \mid P_a(X_i))$$

Here, $P_a(X_i)$ denotes the set of parent nodes of X_i in the graph. The number of parameters required to characterize the joint distribution is significantly reduced by this factorization, which makes use of the conditional independence attributes inherent in the network's topology.

Types of Probabilistic Models

1. **Finite Mixture Models (FMMs)**

   A dataset can be expressed as a combination of several separate distributions, each of which represents a different subpopulation within the data, using probabilistic models known as finite mixture models. The overall model is defined as the weighted total of the components of the mixture model, each of which represents a distinct distribution.

   The probability density function (PDF) of a finite mixture model can be mathematically expressed as:

$$P(X) = \sum_{k=1}^{K} \pi_k P_k(X)$$

where, X is the observed data, K is the number of components in the mixture, π_k is the weight or mixing coefficient for the k-th component (where $\sum_{k=1}^{K} \pi_k = 1$), $P_k(X)$ is the probability distribution of the k-th component.

Applications in NLP

For tasks involving the identification of latent structures in the data, such as: Finite Mixture Models are especially helpful.

- **Topic Modeling:** By considering each topic as a distinct distribution across words, FMMs are able to recognize discrete themes within a corpus. The model efficiently clusters documents into topic groups by allocating weights to each subject according to their significance to particular documents.

- **Clustering**: By simulating the underlying distributions of characteristics (such as word frequencies) that define documents, FMMs can assist in the grouping of related documents in text clustering applications.

2. **Hidden Markov Models (HMMs)**

   HMMs are a class of probabilistic models that capture temporal dependencies and are intended to handle sequential data. The underlying assumption of HMMs is that the system under study is a Markov process with hidden (unobserved) states that change over time in response to observable outputs. An HMM is characterized by:

   - A set of hidden states $S = \{S_1, S_2, \ldots, S_N\}$
   - A set of observable outputs (emissions) $O = \{O_1, O_2, \ldots, O_T\}$
   - Transition probabilities $A = P(S_t \mid S_{t-1})$ indicating the likelihood of moving from one hidden state to another.
   - Emission probabilities $B = P(O_t \mid S_t)$ that define the likelihood of observing a specific output from a given hidden state.
   - An initial state distribution $\pi = P(S_1)$ that defines the probability of starting in each hidden state.

Applications in NLP

- **Part-of-Speech Tagging**: Using the words that come before them, HMMs can model the order of words in a phrase and assign each word to its most likely part of speech. The words in the phrase are the observable outputs, whereas the concealed states are the speech portions.
- **Speech Recognition:** In order to capture the temporal dynamics of spoken words, HMMs are essential to speech processing. The visible outputs correlate to auditory signals, whereas the hidden states represent various phonemes or words.

Posterior Predictive Distributions

Given the posterior distribution over the parameters θ, utilize the information to predict a new output y^* for a given input x^*. This is achieved by marginalizing over the possible values of the model parameters:

$$p\left(y^* \mid x^* \mid, D_{abs}\right) = \int_{\theta} p\left(y^* \mid x^* \mid, \theta\right) \cdot p\left(\theta \mid D_{abs}\right) d\theta$$

Several critical factors influence the effectiveness of posterior predictive distributions:

- Relies on proper model specification (true $\theta \in$ model family θ)
- Requires a right prior distribution ($p(\Theta)$ is accurate)
- Quality of posterior approximation affects accuracy (marginalizing over θ)

Fuzzy Logic

A strong mathematical foundation for handling ambiguity and imprecision in language is provided by fuzzy logic, which allows systems to make conclusions based on approximations rather than true-or-false assessments. This method is particularly useful in Natural Language Processing (NLP), where precise matches are sometimes impossible due to the ambiguity and variety of human language. Fuzzy logic can handle ambiguity in language by permitting degrees of truth instead of rigid boolean logic, which makes it ideal for jobs like text categorization, information extraction, and sentiment analysis.

A proposition is either true (1) or false (0) in conventional logic systems. However, by adding a membership function that converts every given input to a value between 0 and 1, fuzzy logic expands on this binary method. This number indicates how much an element belongs to a particular set. This gives NLP the ability to assess how well a word, phrase, or feeling fits into pre-established language categories. For example, using conventional sentiment analysis algorithms, a term like "pretty good" may not cleanly fall into a "positive" or "negative" category when evaluating the sentiment of a document. A more detailed comprehension of sentiment may be achieved by using fuzzy logic to instead give the sentence a degree of positivity (for example, 0.7 on a scale of 0–1).

In NLP applications that need to deal with polysemy (words with many meanings) and synonymy (different words with similar meanings), fuzzy logic's capacity to approximate reasoning based on contextual information is very helpful. In these situations, ideas with overlapping properties can be represented using fuzzy sets:

- *Word Ambiguity Resolution*: Fuzzy logic may use syntactic and semantic information to identify the most likely interpretation of phrases that have many meanings depending on the context. The term "cold" might, for instance, describe a low temperature or a disease. The likelihood of each interpretation might be weighed by dynamically adjusting the fuzzy membership functions according to the surrounding words.
- *Qualifiers and Linguistic Hedging:* The effects of qualifiers like "very," "slightly," or "almost" can be modeled using fuzzy logic. By modifying the membership values in accordance with the modifier's intensity, fuzzy logic enables these terms—which by their very nature add ambiguity into language—to be included into reasoning processes.

Fuzzy Inference Systems for NLP

Fuzzy logic principles are used by a fuzzy inference system (FIS) to map inputs (linguistic phrases, semantic characteristics, etc.) to outputs (classification categories, sentiment ratings, etc.). Several crucial phases are involved in the process:

- *Fuzzification:* Fuzzification is the process of converting input data—such as words or phrases—into fuzzy sets with matching membership functions. The fuzzification method for sentiment analysis may entail assigning fuzzy categories with different levels of membership, such as "positive," "neutral," or "negative," to emotive terms.

- *Rule Evaluation*: The fuzzified data is subjected to predefined fuzzy rules that represent linguistic expertise. A rule may say, for instance, that if a statement has a lot of "positive" words and few "negative" ones, the attitude should be categorized as "positive."
- *Aggregation of Results:* The system's comprehension of the input is represented by the final fuzzy output, which is created by combining the fuzzy outputs from each rule.
- *Defuzzification*: It is the process of turning the combined fuzzy result into a clear output, such a sentiment score, which may then be utilized for additional processing or decision-making.

5.3 Innovative Machine Learning Models for NLP

By enabling data-driven approaches that can identify patterns and draw conclusions from massive textual datasets, machine learning has completely transformed natural language processing (NLP). This change has opened the door for advanced techniques that go beyond conventional rule-based strategies, allowing systems to more successfully manage the ambiguities and complexity of natural language. This section examines state-of-the-art machine learning approaches that deal with text analysis uncertainty, emphasizing deep learning models and transfer learning methodologies. These methods improve the resilience and flexibility of NLP systems, offering complex ways to comprehend and process text in a variety of difficult situations.

5.3.1 Deep Learning for Uncertainty in Text Analysis

By using complex neural architectures like transformers, recurrent neural networks (RNNs), and convolutional neural networks (CNNs) to address the inherent uncertainty and unpredictability in text input, deep learning has greatly increased natural language processing (NLP). Machine translation, sentiment analysis, text summarization, named entity recognition (NER), and question answering are just a few of the NLP tasks that these models may excel at because of their capacity to learn and represent intricate linguistic relationships and structures.

Bayesian Networks

Graphical models that represent probabilistic interactions between a collection of random variables are called Bayesian Networks (BNs), often referred to as belief networks or probabilistic Directed Acyclic Graphs (DAGs). They are made up of directed edges that show the conditional dependencies between the variables and nodes that represent the variables.

When dealing with dynamic and ambiguous language, Bayesian inference makes it possible to continuously update probability estimates as new information becomes available. The Bayes theorem is used to compute the posterior distribution $(H|E)$ upon the observation of fresh evidence E:

$$P(H|E) = \frac{P(E|H)P(H)}{P(E)}$$

where H represents a hypothesis or model parameter, P(H) is the prior probability, P(E|H) is the likelihood, and P(E) is the marginal likelihood. In tasks like part-of-speech tagging, word meaning disambiguation, and syntactic parsing, Bayesian inference improves accuracy by allowing model predictions to be refined depending on new information.

Conditional Independence and Factorization

The primary benefit of Bayesian networks is their ability to include conditional independence connections between variables, which significantly streamlines the representation of the joint distribution. For instance, if X_1 is independent of X_3 given X_2, the network's topology can explicitly represent this conditional independence, resulting in a reduced factorization. Examine the following three variables: A, B, and C. A influences B, and B influences C. Their joint probability would be factorized as follows:

$$P(A,B,C) = P(A) \times P(B|A) \times P(C|B)$$

Depending on how many parents each variable has, this factorization drastically reduces the number of parameters required to express the joint distribution from 2^n (where n is the number of variables) to a significantly smaller number. For both accurate and approximate inference, a variety of methods are employed, each with unique trade-offs between scalability and computing efficiency.

Bayesian Active Learning by Disagreement (BALD)

Bayesian Active Learning by Disagreement (BALD) is a strategy within active learning that focuses on selecting the most informative unlabeled data points to label next. The goal is to reduce epistemic uncertainty, which arises from a lack of knowledge about the model itself, by iteratively refining the model through additional labeled data. However, it's important to note that acquiring more training data only helps in reducing epistemic uncertainty and does not address aleatoric uncertainty, which is inherent in the data itself.

The BALD approach, introduced by [8], targets data points that maximize the information gain. Specifically, it selects the points where the model exhibits high epistemic uncertainty by identifying instances where different model samples (generated through techniques such as Bayesian sampling or dropout) produce confidently diverging predictions.

$$H\left[y|,x|,D\right]-E_{\theta\sim p(\theta|D)}\left[H\left[y|,x|,\theta\right]\right]$$

where, $H[y|x,D]$ is the overall entropy of model predictions; $E_{\Theta \sim p(\Theta|D)}[H[y|x,\Theta]]$ is the average entropy across samples of model parameters (or dropout masks).

Conformal Prediction

Conformal prediction is a statistical method used to generate prediction sets with a guaranteed level of confidence [9]. It provides a systematic approach to uncertainty quantification for machine learning models, ensuring that the prediction sets produced are likely to contain the true outcome with a specified probability.

Framework Overview

1. Problem Setup: Given a dataset of n exchangeable examples (X_1,Y_1), (X_2,Y_2),…, (X_n,Y_n), where X_i represents the input features and Y_i represents the corresponding label or target value.
2. The goal is to make predictions for a new input, X_{n+1}, while quantifying the uncertainty associated with this prediction.
3. The framework aims to produce a prediction set $C_n(X_{n+1})$ that covers the correct label Y_{n+1} with a specified probability.

Significance Level

- The desired significance level is denoted by ϵ, where $0 < \epsilon < 1$.

- This parameter determines the level of confidence for the prediction set. Specifically, $1 - \epsilon$ represents the coverage probability, which is the likelihood that the true label will fall within the prediction set.

Validity of the Predictor

- A conformal predictor is considered valid if the prediction set $C_n(X_{n+1})$ covers the true label Y_{n+1} with a probability of at least $1 - \epsilon$:

$$P\left(Y_{n+1} \epsilon C_n\left(X_{n+1}\right)\right) \geq 1-\epsilon$$

- This means that, over multiple instances, the prediction set should contain the true label at least $1 - \epsilon$ of the time, providing a guarantee on the model's reliability.

Efficient Conformal Predictor

$$E\left[\left|C_n\left(X_{n+1}\right)\right|\right] \ll |y|$$

Attention Mechanisms in Transformers

Techniques for Attention-Based Uncertainty Management

In NLP tasks, attention processes also provide methods for managing and quantifying uncertainty:

- **Attention Weights Analysis:** We may decipher the words or phrases the model deemed most important for a particular prediction by looking at the attention weights. In situations when comprehending the reasoning process is crucial, this might aid in diagnosing the reasons behind a model's decisions, hence ensuring explainability.
- **Attention Masking:** Some tasks may contain input that is less instructive or even deceptive. By limiting the model's focus to more pertinent textual segments, attention masking approaches can help lower output noise and uncertainty.
- **Hierarchical Attention Networks**: Hierarchical attention models may be used for materials that have several levels of organization, such as words, phrases, and paragraphs. These models reduce uncertainty across several granularities by applying attention at distinct levels, allowing them to capture more abstract associations at the sentence or paragraph level and fine-grained information at the word level.

Extensions of Attention Mechanisms for Enhanced Performance

A number of advanced techniques have been created to further exploit the capabilities of attention mechanisms in handling uncertainty:

- **Memory-Augmented Attention:** Attention mechanisms can make use of extra contextual information that might not be included in the current input sequence by incorporating external memory into the model. This is helpful in situations when pertinent information may be found in a different knowledge base, such as when addressing open-domain questions.
- **Adaptive Attention Span:** Models with adaptive attention span dynamically modify the range of their attention depending on the type of input, as opposed to employing a fixed attention window for every token. For example, shorter sequences can employ a smaller attention span for more focused attention, whereas longer sequences may benefit from a broader span to preserve context over longer text parts.
- **Sparse Attention Mechanisms:** Sparse attention approaches can restrict the amount of tokens taken into consideration during attention computation when only a small portion of the input is actually important. By concentrating on crucial material, this efficiently manages uncertainty while lowering computing complexity while maintaining the most crucial contextual information.

5.3.2 Transfer Learning in Sector-Specific NLP Applications

By allowing the application of pre-trained models from general-purpose language representations to specialized, domain-specific tasks, transfer learning has completely transformed natural language processing. In order to overcome uncertainty in text analysis, particularly when working with sparse domain-specific data, this method enables models to exploit significant language comprehension obtained from large-scale datasets.

Pre-training Phase

Pre-training a language model on a sizable, varied corpus is the initial stage in transfer learning. Commonly utilized models are RoBERTa (Robustly Optimized BERT), GPT (Generative Pre-trained Transformer), and BERT (Bidirectional Encoder Representations from Transformers). Pre-training assignments frequently consist of:

- **Masked Language Modeling (MLM)**: Random words in a phrase are masked in models such as BERT, and the model is trained to predict the masked words. This aids in the model's acquisition of word context by capturing syntactic and semantic links.
- **Next Sentence Prediction (NSP)**: The model learns discourse-level relationships by determining if a given phrase logically follows another.

- **Casual Language Modeling (CLM)**: In models like GPT, Causal Language Modeling (CLM) encourages the model to comprehend forward dependencies in the text by training it to anticipate the next word in a sequence.

Fine-Tuning Phase

- After pre-training, a smaller, domain-specific dataset customized for a specific task—like sentiment analysis, entity recognition, or question answering—is used to fine-tune the model.
- While maintaining the broad linguistic knowledge learned during the pre-training stage, fine-tuning modifies the pre-trained model's weights. The model can become an expert in the target domain's particular vocabulary, syntax, and semantics thanks to this adaptation process.
- Catastrophic forgetting, in which the model loses its general language abilities, can be avoided by employing strategies like layer-wise learning rate modification and gradual unfreezing, which involves training the latter layers first and gradually improving the previous levels.

Technical Considerations in Transfer Learning for NLP

1. Domain Adaptation Techniques: Direct fine-tuning may not produce the best outcomes when the pre-training and fine-tuning domains differ greatly because of a lack of domain alignment. Techniques such as:
   - *Domain-Adaptive Pre-training (DAPT)*: Before fine-tuning on a smaller labeled dataset, the model is further pre-trained on a domain-specific corpus (such as scientific articles).
   - *Multi-task learning* is the process of simultaneously improving the model on several related tasks in order to capture common domain-specific characteristics and enhance generalization.
2. Mitigating Catastrophic Forgetting: General linguistic information learned during pre-training may be lost when a model is fine-tuned for a specific task. Strategies to address this include:
   - *Gradual Unfreezing:* This method involves first fine-tuning the last layers of the model and then gradually unfreezing the previous levels.
   - *Regularization Techniques:* Such as weight decay and dropout, which help prevent overfitting to the fine-tuning dataset.
   - *Continual Learning Approaches:* In order to maintain general language understanding, the model is frequently retrained on general datasets.

3. Performance Optimization and Hyperparameter Tuning:
    - Fine-tuning a pre-trained model requires careful hyperparameter tuning, including learning rate schedules, batch size adjustments, and early stopping to avoid overfitting.
    - *Layer-wise Learning Rate Decay (LLRD):* Applying different learning rates to different layers, typically assigning lower learning rates to the base layers (closer to the input) and higher learning rates to the task-specific layers.

5.3.3 Few-Shot and Zero-Shot Learning for Uncertainty Reduction

In machine learning and natural language processing (NLP), few-shot and zero-shot learning are advanced techniques that address situations in which there is limited to no task-specific training data available. These methods are particularly useful for dealing with text analysis uncertainty because they allow models to execute new tasks or adjust to new domains with little data. They improve NLP programs' adaptability and effectiveness, particularly in fields where data annotation is costly or time-consuming.

Few-Shot Learning

Few-shot learning involves training a model to execute a new task by generalizing from a limited number of labeled instances (e.g., 1–10 samples). Utilizing past information gained from extensive pre-training on general-purpose datasets, this is achieved. Among the few-shot learning strategies are:

- *Meta-Learning*: Also referred to as "learning to learn," meta-learning allows models to swiftly adjust to new tasks by optimizing for generalization across tasks through training on a range of activities. With this method, a meta-learner modifies the main model's parameters to enable it to adapt to novel jobs with less data.
- *Prompt-Based Fine-Tuning:* Prompt engineering, in which a few instances are given in the input prompt to direct the model in carrying out the intended job, can be used to accomplish few-shot learning for models such as GPT-3 and T5. From these instances, the model draws conclusions and applies them to novel, invisible inputs.
- *Prototypical Networks and Metric Learning*: These techniques classify new examples based on their similarity to a small set of "prototype" representations for each class, reducing the need for large training datasets by leveraging learned distance metrics.

Zero-Shot Learning

- By using pre-existing information from its pre-training phase, zero-shot learning allows a model to complete a task without the need for task-specific training instances.
- The model matches inputs to classes based on label semantics or natural language descriptions using semantic embeddings, such as word or sentence embeddings.
- Among the methods for zero-shot learning are:
  - *Natural Language Inference (NLI):* The target goal is framed as an inference issue, asking the model to determine if a certain input (such as "The sentiment of this review is positive") indicates a task-specific description.
  - *Label Embeddings and Matching*: Zero-shot models are able to determine relations between inputs and class labels according to their semantic closeness by mapping class labels or descriptions into a common embedding space.
  - *Prompt-Based Approaches:* Models such as GPT-3 can be directly instructed to perform a task using task-specific prompts without explicit fine-tuning, leveraging their broad understanding of language acquired during pre-training.

5.4 Conclusion

Managing uncertainty in NLP necessitates a multifaceted approach that combines domain-specific knowledge, expert systems, and cutting-edge machine learning models. Knowledge-based systems provide a framework of structured rules and established heuristics, which are particularly effective for deterministic tasks such as parsing structured text and extracting specific information. These systems enhance accuracy by leveraging domain knowledge to resolve ambiguities inherent in natural language, ensuring that the context and nuances of the language are properly interpreted.

On the other hand, innovative machine learning models tackle uncertainty by learning patterns from vast datasets. Techniques such as deep learning enable these models to capture intricate language features and adapt to various domains, allowing them to address ambiguities and variations in language use. By integrating statistical approaches with rule-based systems, NLP applications can benefit from both the flexibility of data-driven methods and the precision of expert knowledge.

The integration of these methodologies leads to improved accuracy and reliability across a range of NLP applications, including sentiment analysis, machine translation, and information retrieval. However, the field remains in a state of flux, with ongoing research required to tackle persistent challenges. Issues such as knowledge acquisition bottlenecks, the dynamic nature of domain-specific knowledge, and the complexities of model generalization in specialized contexts must be addressed to fully harness the potential of knowledge integration in NLP systems. As the landscape evolves, continued exploration of these challenges will be crucial for advancing the capabilities of NLP technologies.

References

1. Xiao Y., & Wang W. Y. (2019, July). Quantifying uncertainties in natural language processing tasks. In *Proceedings of the AAAI conference on artificial intelligence* (Vol. 33, No. 01, pp. 7322-7329).
2. Ulmer D. (2024). On Uncertainty In Natural Language Processing. *arXiv preprint arXiv:2410.03446.*
3. Ulmer D., Frellsen J., & Hardmeier C. (2022). Exploring predictive uncertainty and calibration in NLP: A study on the impact of method & data scarcity. *arXiv preprint arXiv:2210.15452.*
4. He J., Zhang X., Lei S., Chen Z., Chen F., Alhamadani A., & Lu C. (2020, November). Towards more accurate uncertainty estimation in text classification. In *Proceedings of the 2020 Conference on Empirical Methods in Natural Language Processing (EMNLP)* (pp. 8362-8372).
5. Jean P. A., Harispe S., Ranwez S., Bellot P., & Montmain J. (2016, June). Uncertainty detection in natural language: A probabilistic model. In *Proceedings of the 6th International Conference on Web Intelligence, Mining and Semantics* (pp. 1-10).
6. Campos M., Farinhas A., Zerva C., Figueiredo M. A., & Martins A. F. (2024). Conformal prediction for natural language processing: A survey. *Transactions of the Association for Computational Linguistics, 12*, 1497-1516.
7. Beck C., Booth H., El-Assady M., & Butt M. (2020, December). Representation problems in linguistic annotations: Ambiguity, variation, uncertainty, error and bias. In Proceedings of the 14th Linguistic Annotation Workshop (pp. 60-73).
8. Houlsby, N., Huszár, F., Ghahramani, Z., & Lengyel, M. (2011). Bayesian active learning for classification and preference learning. *arXiv preprint arXiv:1112.5745.*
9. Zhou X., Liu H., Pourpanah F., Zeng T., & Wang X. (2022). A survey on epistemic (model) uncertainty in supervised learning: Recent advances and applications. Neurocomputing, 489, 449-465.

Chapter 6
NLP for Fraud Detection and Security in Financial Documents

Shobha Bhatt and Geetanjali Garg

Abstract Natural language processing (NLP) has impacted the area of information security in different ways. NLP is widely used to detect anomalies and fraud in financial documents. Different NLP techniques such as named entity recognition (NER), sentiment analysis, text classification, word embedding, keyword extraction and similarity detection are applied to detect financial document fraud. Further large language models (LLM have also opened a different era of security mechanisms by providing comprehensive and reasoning-based outputs. This chapter aims to highlight different anomaly and fraud detection techniques using NLP. Other security measures which can be taken by using NLP are also discussed. Further integration of machine learning and deep learning-based algorithms with NLP is also described to enhance security in financial documents of large structured and unstructured data sets. Different supervised, unsupervised, and semi-supervised machine learning models are being applied to detect anomalies and fraud in financial documents by learning from past data. NLP-based implementation using machine learning is also elaborated. Case studies are presented to gain insights into NLP in detecting financial fraud. Finally, lessons learnt are summarized with a conclusion.

S. Bhatt (✉)
Department of Computer Science and Engineering, Netaji Subhas University of Technology, East Campus, New Delhi, Delhi, India
e-mail: shobha.bhatt@nsut.ac.in

G. Garg
Department of Software Engineering, Delhi Technological University, New Delhi, Delhi, India
e-mail: geetanjali.garg@dtu.ac.in

A. Kumar, S. R. Sangwan (eds.), *Transformative Natural Language Processing*,
https://doi.org/10.1007/978-3-031-88988-2_6

Recently, with vast online processing and networked environments, financial security has faced losses due to different frauds and anomalies. Further dependence on technological advancements to safeguard against fraud and other financial crimes is the need of the hour. Natural Language Processing (NLP) has played a major role in anomaly and fraud detection. By applying computational linguistics, NLP techniques process vast amounts of structured and unstructured textual data, to identify patterns that may lead to different fraudulent activities. NLP-based security techniques for anomalies and fraud detection within time can stop financial losses to enhance customer trust and ensure regulatory compliance.

Different NLP-based methods are being applied for anomaly detection. Anomaly is an act which deviates from normal behavior. Text classification [1], named entity recognition [2], and sentiment analysis [3], are being experimented with anomaly detection. These techniques are applied to various types of financial documents such as emails and log files using NLP methods. Word embeddings [4] and Latent Semantic Analysis (LSA) [5] have the power to understand semantic relationships within textual datasets to detect hidden patterns for fraudulent activity. Nowadays, an amalgamation of NLP and artificial intelligence is also being applied to take advantage of artificial intelligence to secure financial documents. Integration of NLP and machine learning techniques allows systems to learn from past fraudulent behaviors and adapt to new, emerging patterns. These models are classified into supervised, unsupervised, [6] and semi-supervised approaches. Supervised models are trained on labeled datasets of both fraudulent and legitimate transactions and classify the data into different categories. Unsupervised models [7] effectively identify new fraud patterns that have not been previously labelled, thus identifying new types of fraud and anomalies. Semi-supervised models combine aspects of both, utilizing limited labelled data alongside larger amounts of unlabeled data to enhance detection capabilities.

Behavioral analysis is also being experimented with to detect anomalies in large financial documents. For analyzing behavior for fraudulent activities unusual keyword usage, changes in sentence structure, and emotional shifts in tone are recorded. Further frequency analysis, intent analysis, and linguistic profiling help understand deviation from normal patterns. This multi-dimensional approach offers a robust framework for fraud detection, combining textual content analysis and behavioral indicators to improve system reliability and reduce false positives.

Advanced language models such as Bidirectional Encoder Representations from Transformers (BERT) and Generative Pretrained Transformers (GPT-3), along with advanced NLP techniques are transforming financial security systems. These models capture and understand hidden patterns within the text to enhance the precision of fraud detection systems and apply security measures for both financial institutions and their clients. The expansion of NLP technologies has opened different avenues for fraud and anomaly detection in financial documents. This chapter will explore various techniques, case studies, and lessons learned in the field, elaborating on how NLP is reshaping anomaly detection and fraud prevention in the financial sector.

6.1 Techniques for Anomaly and Fraud Detection Using NLP

Different techniques and methods are being applied to detect anomalies and fraud in financial documents. Anomaly detection is the process of identifying deviations from normal data patterns and behavior. Anomalies can be detected by applying different NLP techniques to the dataset. The researchers have applied techniques like text summarization and word embedding to detect anomalies. The text summarization technique is applied to the logfile to understand and make decisions in the process of abnormal events while word embedding techniques are applied for finding the semantic description from available data for anomaly detection [8].

Automated fraud detection has played an important role in recent years in the early prevention of fraud. The use of natural language processing techniques and tools has been successful towards achieving this. The tool Konstanz Information Miner (KNIME) has been widely used. It allows text mining on unstructured data without much effort in coding. Latent Semantic Analysis (LSA) was used to detect patterns in documents related to financial fraud. Other techniques used are the Bag of Words (BOW), term frequency and inverse term frequency. BOW converts text into numerical data by counting word occurrences so that further processing can be done for fraud detection [9]. The next subsection highlights techniques for identifying fraud detection and security measures.

6.1.1 NLP Methods for Identifying Fraud in Transactions

Different NLP methods such as text clustering, text classification, word embedding, named entity recognition, sentimental analysis, contextual language models, text similarity detection, keyword spotting and other related techniques are applied for identifying fraud in transactions. NLP methods for fraud detection can be summarized as shown in Fig. 6.1. Further subsections briefly explain the different techniques.

- **Text Classification:** It is used to classify text data into binary classes such as fraudulent or non-fraudulent. Researchers have implemented text classification to identify spam mail and ham mail by using NLP techniques combined with machine learning algorithms. The main steps are as follows. Initially, the related dataset is loaded, and the contents of text files are read. After that, the dataset is pre-processed by using NLP methods such as tokenization, stop word removal, deleting numbers, punctuation and stemming. Tokenization divides whole documents into small units so that different processing can be applied. The stop word removal process removes words such as "the", and "is" which are not significant for further processing. Stemming is the process of converting the word to its root format so that information retrieval and other processing tasks can be applied.

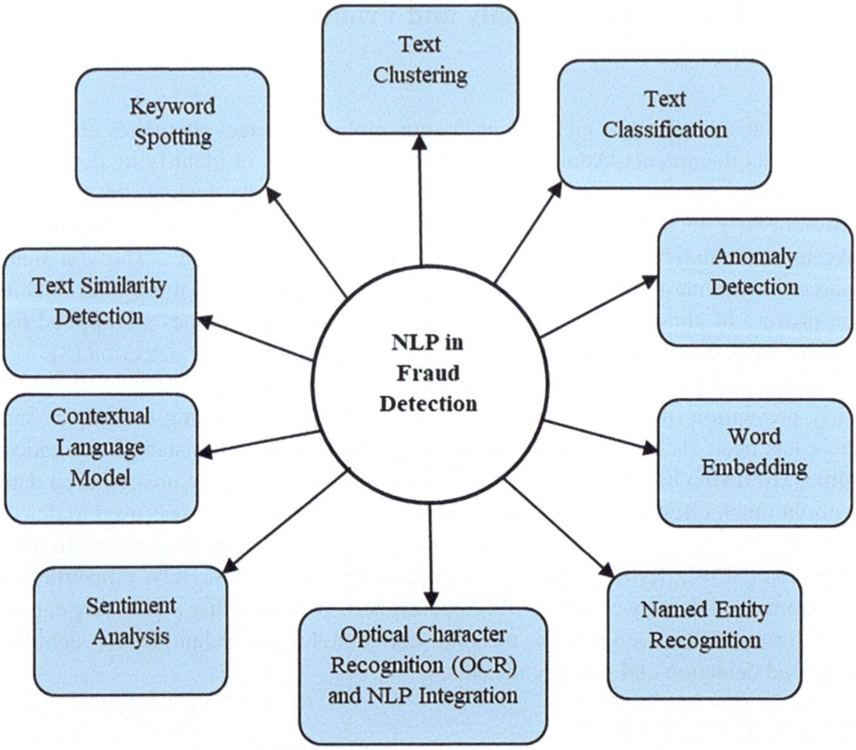

Fig. 6.1 NLP techniques for fraud detection

Similarly, numbers and punctuation marks can also be removed when they do not make meaningful contributions. After preprocessing features are extracted and after feature extraction, the feature set is divided into training and testing data sets to apply different machine learning algorithms that classify mail into spam and ham. Lastly, a confusion matrix is generated after classification [10].

- **Example Use Case**: Fraud detection was experimented on the available Bkash-related scam data. Bkash is a well-known financial transaction service in Bangladesh. The data was collected from different websites, Facebook, YouTube and different blogs where an individual used to express their experiences. Afterwards, data collection data was labeled as fraud and partial fraud. The fraud label is related to the event when the attacker has performed the fraud, and the partial fraud label is when the attacker tries to commit fraud but fails. Preprocessing steps using NLP such as elimination of punctuation, removal of duplicate data, removal of short-length sentences with no meaning, stop word removal, stemming, tokenization and word-to-vector conversion using TF-IDF were applied. Finally for text classification different learning algorithms were applied. It was revealed that the most effective classifier was the stochastic gradient descent algorithm with 77.78% accuracy [11].

- **Named Entity Recognition (NER):** This process is applied to identify entities like people or organizations and other relevant factors in textual documents. Researchers applied different tools for named entity recognition for different documents including financial documents. The data set consisting of 47,959 sentences and 1,048,575 words from the Kaggle repository was applied for different named entity recognition. Different tools such as Azure Cognitive Service Tool, Stanford Core NLP, Spacy and NLTK were used for entity recognition. The entities related to geography, geopolitical, organization, person, time indicator, artefact and natural phenomenon were used. It was revealed that the Azure Cognitive Service Tool provided the best results for named entity recognition among the different tools. Bidirectional Encoder Representation from a transformer-based model also provides better results [12].
- **Example Use Case**: NER can be applied to detect fake companies and vendors by extracting information such as the name of the company, address, and other financial details from the billing and any other related documents. The extracted information of the vendors from the fake document can be matched with the actual vendor to detect fraudulent activity.
- **Sentiment Analysis:** This is the process of recognizing the tone of part of the text. The sentiment analysis can be applied to determine fraudulent behavior. The main challenge faced during sentiment analysis for financial documents is the non-availability of large, labelled databases and imbalanced datasets. The research work has been conducted by applying NLP-based sentiment analysis starting from a lexicon-based approach to recent NLP transformers. It was concluded that NLP-based transformers are providing better results [13].
- **Example Use Case**: Sentimental analysis was performed using data from different companies. The data was extracted from PDF documents of three companies' financial case studies and cleaned. Sentiment analysis was performed by using the TextBlob natural language processing pre-trained model. Two matrices, polarity and subjectivity were used for the performance analysis [14].
- **Topic Modeling:** It is the process of finding common topics from the documents. Different patterns can be detected in fraud-related text by analyzing the topics for fraudulent activities. Researchers have applied NLP-based methods to divide the documents into different topics. The Latent Dirichlet Allocation (LDA) technique is commonly used for topic modelling. This technique is based on the concept that documents can be divided into different topics [15].
- **Example Use Case**: The knowledge base for fraud was created to bridge the gap between the financial sector and IT by using topic modelling. Initially, the data was collected regarding different frauds from websites such as the Times of India, Yahoo, Hindustan Times and RBI. The key phrase identification was performed using TF-IDF. Topic modeling was applied on key phrases using LDA with the MALLET package in Python. Finally, knowledge base ontology was created using Wordnet and Protégé tool. Knowledge-based ontology can be applied to detect new fraud [16].
- **Anomaly Detection:** It is the process of finding unusual patterns in financial documents. It involves identifying deviations from typical or expected data pat-

terns, which is crucial in fraud detection. LLMs can analyze sequential data for detecting unusual patterns benefiting financial industries for anomaly detection. LLMs are capable of processing and analyzing temporal dependency in large sets of data. BERT models are also being applied for anomaly detection to refer to long-term dependencies. LLMs can also scrutinize log data. Due to the ability of the LLMs to understand contextual data they can classify and respond to abnormal patterns in the system. BERT logs are in use as the log sequences are considered NLP sequences for further processing. ChatGPT can also be asked to understand log files by creating appropriate prompts. Term Frequency-Inverse Document Frequency (TF-IDF) based features are also being applied [17].

- **Example Use Case**: In online banking, the process starts with the data collection of transaction details like amount, time, geographical site, user behavior patterns such as login history, system details and previous fraud cases. Preprocessing is applied to remove outliers after data collection to extract knowledge for model understanding. Then, the Machine Learning model is trained to identify the suspected activity from normal ones. Whenever the system detects any unexpected behavior, it automatically blocks the transaction and feedback to evolve for future frauds.
- **Word Embeddings:** It is the technique of NLP which is used to convert sentences or words into numeric values to understand semantic meaning and the relationship between the words. Research work was conducted to detect fraud in Chinese financial documents. First, the Chinese word segmentation was applied then stop words were removed according to the stop word listed for financial documents. After that word embedding techniques are applied for inputting the data into predictive models. Different predictive models such as extreme gradient boost (XGB), random forest (RF), support vector machine (SVM), artificial neural network (ANN) and convolutional neural network (CNN) were applied. It was revealed that CNN and RF performed better than other models [18].
- **Example Use Case**: Banking information is unstructured and problematic for analysis directly. So, to resolve this, the word embedding technique transforms the details into numeric representations to get the semantic meaning of words and phrases. The process starts with the removal of irrelevant stop words and applying tokenization. Then, these words are fed into various Machine Learning models to discover hidden patterns and unusual patterns.
- **Keyword Spotting**: It identifies specific trigger words or phrases commonly associated with fraudulent activities and examines deviations from typical behavior or patterns. Using this technique, transaction descriptions, or user interactions are scanned for keywords or phrases that might indicate fraud [19, 20]
- **Example Use Case**: If a customer reports triggering words in communication like a stolen pin or emergency overseas transfer, the system marks them as suspicious interactions and injects these words into the keywords list as a future preventive measure. The process initializes with the updating of keyword spotting system to automatically raise a flag while encountering any prompt keywords.
- **Text Clustering**: This process groups similar documents or text based on linguistic patterns to detect recurring fraud schemes. Researchers have applied

K-means clustering to detect fraud in financial documents. By clustering vast amounts of financial data fraudulent patterns and behavior can be identified on time and resources can be utilized by focused monitoring [21].

- **Example Use Case**: In an insurance company, events like unforeseen accidents and ambiguous medical claims are clustered, which represents similar language patterns. Such recurring activities are grouped which makes the system alert and raise flags. After such data collection, the preprocessing involves extracting these data and applying K-means clustering to quickly identify indicative patterns across huge data. This overall reduces the risk of financial loss and raises fraud detection accuracy.
- **Optical Character Recognition (OCR):** Optical character recognition combined with NLP can be used to analyze scanned documents and extract textual data for fraud detection in financial reports, contracts, or invoices. Researchers applied OCR technology in anti-money laundering (AML) systems. OCR technology plays an important role in automating AML systems, thus improving accuracy, and protecting against money laundering [22].
- **Example Use Case**: A global bank applies this technique to scan invoices from large volumes of scanned physical documents and transform them into machine-readable text. After data collection, the system extracts the main data to process it further through the NLP algorithm. It then finds patterns and links that present money laundering tasks. Later, the suspicious documents route towards the AML team for further investigation.
- **Text Similarity Detection:** It compares new texts with known fraudulent documents to identify suspicious patterns. The degree of similarity between the document vectors that correspond to fraud reports is being measured using several techniques, such as Cosine similarity and Jaccard similarity. This approach allows for quicker identification of fraud patterns and improves fraud prevention efforts [23].
- **Example Use Case**: A new loan application consists of language like the past fraud reports. The data collection process gathers such data and after preprocessing, the Cosine similarity or Jaccard similarity is applied to compare vectors of novice application with the previously flagged fraudulent report. These similarity vector scores define the potential to flag the report as fraudulent or not.
- **Language Models:** Language models like large language models (LLMs) are being trained on massive text data to generate human-like texts. Different models like BERT and GPT-3 enhance fraud detection by understanding the meaning and context of words in customer interactions, emails, and chat logs. These models can be fine-tuned to understand domain-specific language understanding and contextual meaning in finance-related tasks [24].

- **Example Use Case**: The study was presented for detecting scams or legitimate mail by using large language models. The data was collected from different emails. The data was cleaned by removing irrelevant information, formatting the text, tokenization and stemming. In the next step, data was labelled as scam or legitimate. After that models GPT 3.5 and GPT 4 were applied to detect fraud.

6.1.2 Security Measures and NLP

Security measures using NLP refer to applying actions, methods, and planning to protect and secure the data. GPT-4 large language model, BERT and RoBERTa NLP models with fine-tuning were applied for spam classification. Experimental results show that the GPT-4 based model outperformed compared to other models by using LLM-based models, the limitation of traditional spam detection models can be overcome in case of increased threats [25]. NLP-based measures can also be applied to de-identify personal information from the data. The research work was done to redact student personal data using GPT-4. The data was collected from massive online course discussion forums. It was revealed that the prediction of such redaction was worse for names and locations [26]. NLP-based dynamic passwords are generated using GPT-2 and verification is done using natural language understanding (NLU) based models. For the verification, a set of specific criteria and threshold adjustments were applied by NLU models. This approach is significant in safeguarding the vast amount of data [27]. Defense from jail board prompts can be done by using different NLP-based models by applying NLP-based methods such as content filtering and contextual understanding. The jail board prompts are the commands which are being used by hackers to bypass AI-based ethical guidelines and policies. Attackers use different types of methods such as feeding into large language models with indirect requests, using multiple instructions for combining the output of the instructions, role-playing commands and embedding incorrect requests to deceive the model [28]. Further sections elaborate on different security measures.

- **In-depth analysis of NLP Techniques for Security**
- Dynamic text-based password generation is explained by using pre-trained models as described below.
  (a) **Dynamic Text-Based Security Mechanisms**: The study was presented to generate dynamic passwords. A large data set containing text samples was cleaned and pre-processed using NLP methods. Following that GPT-2 pre-trained model was applied to generate a password. After that generated passwords were fine-tuned to meet security requirements and finally passwords were verified using natural language understanding (NLU).
  (b) **Description of NLP-driven dynamic password generation:** Firstly, input was taken from the user then the user's input was preprocessed. The input pattern was analyzed for linguistic and contextual patterns. Following that dynamic password rules were created, and dynamic password was generated.
  (c) **Overview of verification using NLU model:** The user's input was preprocessed for verification and semantic matching was performed then user's behavioral biometrics data was analyzed, and verification results were returned.
  (d) **Sensitive Information Masking**: Automated redaction of sensitive information using NLP-based entity recognition models was applied. It was revealed that transformer-based trained models can better classify the data

which have contextual effects. BERT (Bidirectional Encoder Representations from Transformers) and RoBERTa (A Robustly Optimized BERT Approach) are both transformer-based NLP models. The main difference between BERT and RoBERTa is given below in Table 6.1 [29].

- **Use of Large Language Models (LLMs) in Security**
- LLMs play an important role in security measures. Different issues such as Fine-Tuning, Transfer Learning and Domain-Specific Adaptations of LLMs are discussed in the following sections.

  (a) **Fine-Tuning and Transfer Learning**: A study was presented to find the usage of LLMs in phishing and advance fee fraud. Different key steps are data collection, preprocessing, training and integration into the target system. With a labeled dataset selected LLM models such as GPT 3.5 and GPT 4.0 were trained and fine-tuned for scam detection. Supervised learning techniques are applied, allowing the model to learn from the labelled examples in the dataset. Finally, models were evaluated to classify text as either a scam or legitimate based on the input labels. Metrics such as accuracy, precision, recall and F1 score were applied to measure the performance of the system [30].

  (b) **Domain-Specific Adaptations of LLMs**: Domain-specific language models address the need for special financial security requirements which general-purpose LLMs might not capture such as industry-specific nuances effectively. The study was presented to fine-tune and utilize domain-specific LLMs. The study starts by collecting datasets related to the financial domain. The dataset includes communication logs, transaction records, and examples of fraudulent activities. The dataset was pre-processed by removing irrelevant information, handling missing values and normalization. The LLMs such as GPT-4 and BERT were applied and fine-tuned. Transfer learning techniques were applied to adapt the pre-trained models to specific financial domain needs. Custom embeddings were applied to better understand the

Table 6.1 Difference between BERT and RoBERTa

	BERT	RoBERTa
Training	Uses Masked Language Model (MLM) and next-sentence prediction	Uses only MLM
Training data	16GB dataset comprising BookCorpus and English Wikipedia	160GB dataset, including additional data sources like Common Crawl News, OpenWebText, and Stories from the BooksCorpus
Training duration and batches	Fewer duration and batches	Larger batch sizes, more epochs, and more input sequences per batch
Parameters	Fewer parameters	Extensively tunes hyperparameters such as learning rate and sequence length to maximize performance

context and terminology used in financial communication. Finally, trained models were evaluated [31].

- **Challenges in Implementing NLP-Based Security Measures**
- Different challenges are faced while implementing NLP based security systems. The following subsection highlights privacy concern, balancing accuracy and privacy, data security and model vulnerabilities.

  (a) **Privacy Concerns**: De-identification techniques remove or change personal and sensitive information from the data. But might struggle with edge cases when they are cross-referenced. For example, the data from the financial documents may be deidentified but it can be reidentified from social accounts when cross-referenced with other data points. If reidentification occurs it may cause financial loss and other privacy breaches. General Data Protection Regulation (GDPR) plays an important role in maintaining data privacy. GDPR directs rigorous guidelines for data processing. These guidelines ensure that organizations use personal data with responsibility and prevent data from reidentification [32].
  (b) **Balancing Accuracy and Privacy**: The study was presented to show the trade-off between data accuracy and privacy. To achieve the balance between these two differential privacy methods adds noise to the actual data. It tries to minimize noise impact on data without compromising privacy [33].
  (c) **Data Security and Model Vulnerabilities**: The study was presented to discuss different attacks on NLP-based models and defenses. The adversarial attacks are performed by using malicious input, gradient-based attacks, semantic attacks and model inversion. The modification to the input may change the output and can generate biased output. The gradient-based attack changes the pathway of model learning and leads to an effect on the model's behavior. Semantic attacks use linguistic knowledge to create deceptive inputs and can cause misclassification. Model inversion attacks take place when the attacker reverses the model and tries to get information. Different prevention techniques such as adversarial training, model regularization, input validation, model transparency, input preprocessing, real-time monitoring and regular retraining are in trend [34].

- **Scalability and Real-Time Processing Challenges**
- NLP-based models have evolved from task-specific to generalized systems to handle different fields. The scalability and real-time processing are major concerns as these models contain trillions of parameters. Different challenges faced by NLP-based models in finance are processing speed, latency, model architecture constraints, scalability Challenges such as volume management, resource allocation, model updates, maintenance, monitoring and logging. The researchers have proposed multilayer architecture, resource optimization, development of faster and more efficient NLP architectures, improved techniques for model compression, dynamic scaling based on load prediction and advanced caching strategies for common patterns [35–39].

- **Monitoring Model Drift:** Model drift is the change in data pattern which leads to a decline in model performance. In recent years fraud patterns have been changing drastically. The attackers are finding new ways to implement fraud. The model drift pattern shows different temporal and behavioral evolution. Temporal patterns include daily pattern shifts, weekly trend changes and seasonal variations and behavioral evolution include changes in transaction amount, frequency of modification and location patterns. The financial framework must be designed to detect new patterns in real time and dynamic adaptation is required to improve fraud detection accuracy and reduce false positives. Some strategies related to model adaptation are continuous model update and ensemble adaptation. The monitoring and updating strategies include real-time monitoring, model update trigger, performance degradation detection, new pattern identification and periodic schedule updates. Other strategies to handle model drifts are incremental learning, continuous model updates, and merging different fraud technologies [40, 41].
- **Ethics and Compliance in NLP-Driven Security Systems**
- NLP driven security system face different challenges due to lack of transparent, bias, explainability, ethical and legal consideration.

  (a) **Ethical Challenges in Automation:**
  The study was presented to address ethical issues in NLP models for Anti-Money Laundering (AML) practices within the financial industry. The challenges are faced due to algorithmic bias caused by training data and these biases can produce false positive results such legitimate transactions can be flagged as fraudulent and can result in customer dissatisfaction and untrust. Further lack of transparency and explainability may lead to a decline in customer trust. Data privacy in NLP-based models in financial sectors is highly sensitive. It needs to be handled carefully. and its misuse can lead to significant ethical and legal considerations. Regulatory standards such as the Payment Card Industry Data Security Standard (PCI-DSS) and the ISO/IEC 2700 must be deployed for the smooth functioning of financial frameworks. PCI-DSS primarily deals with organizations handling payment card data, while ISO/IEC 27001 provides a broader framework for managing information security risks. PCI-DSS ensures secure processing, storage, and transmission of cardholder data. NLP integration must comply with this standard so that financial data is protected against breaches and unauthorized access. The implementation strategies include cryptographic algorithms, access control strategies, regular security audit check-ups, and maintaining a secure infrastructure. ISO/IEC 27001 provides a framework for managing information security risks through an information security management system (ISMS). NLP Integration in financial security must be associated with this standard by applying explainable risk management processes. The implementation involves identifying potential security risks associated with NLP model deployment, implementing controls to mitigate these risks, and continuously monitoring and improving the ISMS [42].

- **Case Studies and Practical Implications**
- In this section, two example case studies are presented on institutions that have implemented NLP-based security measures, such as dynamic password systems and NLP-driven fraud monitoring. The outcomes, benefits, and lessons learned. Also discussed. Comparative analysis of tools such as Azure Cognitive Services, Spacy, and Stanford NLP also presented in security contexts, including speed, accuracy, and ease of integration with other financial security systems.

(a) **Case Study 1: JP Morgan Chase—Streamlining Loan Approvals**

Institution: JP Morgan Chase

Implementation: Advanced AI system for automating loan approval processes.

Working:

1. **Data Collection**: The data was collected from various sources, including applicant financial histories, credit scores, and transaction records.
2. **Data Preprocessing**: The data was cleaned and pre-processed to remove any inconsistencies or irrelevant information.
3. **Model Training**: An NLP-based AI model is trained on historical loan data to learn patterns associated with approved and rejected loans.
4. **Real-Time Processing**: When a new loan application is received, the system processes the data in real time, evaluating the applicant's eligibility based on predefined criteria.
5. **Decision Making**: The AI model decides on the loan application for approval or rejection.
6. **Feedback**: The outcome is communicated to the applicant, and any approved loans are processed further.

Outcomes:

The use of NLP results in increased speed of loan processing, enhanced customer satisfaction due to faster loan approval, and cost efficiency due to reduced manual processes and scalable operations.

Lessons Learned:

- Process efficiency and operational cost reduction are key benefits of AI implementation.
- Ensuring data quality and model accuracy is crucial for reliable decision-making.

(b) **Case Study 2: Bank of America—Erica, the AI-Powered Financial Assistant**

Institution: Bank of America

Implementation: AI-powered financial assistant named Erica.

Working:

1. **User Interaction**: Erica interacts with customers through voice and text, providing personalized financial advice and support.

2. **Natural Language Processing**: Advanced NLP techniques are used to understand and interpret customer queries accurately.
3. **Data Integration**: Erica accesses and analyzes customer financial data, such as account balances, transaction history, and spending patterns.
4. **Personalized Insights**: Based on the analysis, Erica offers personalized insights and recommendations, such as budgeting tips, bill payment reminders, and fraud alerts.
5. **Continuous Learning**: Erica continuously learns from customer interactions to improve its accuracy and relevance over time.

Outcomes:

The outcomes can be summarized as improved customer Service with personalized financial advice and support, enhanced Security with AI-based analysis to detect potential fraud and Operational Efficiency by streamlined customer interactions and reduced workload for human agents.

Lessons Learned:

- AI can enhance customer service and security while improving operational efficiency.
- Continuous learning and adaptation are essential for maintaining the relevance and accuracy of AI-powered assistants [43].

- **Comparative Analysis of Tools**

- These case studies and tools demonstrate the practical applications and benefits of NLP-based security measures in financial systems. By leveraging advanced AI and NLP technologies, institutions can enhance their security, improve customer service, and achieve operational efficiency. Following are few examples of the tools that can be applied for NLP applications.
  - Azure Cognitive Services can be integrated into financial systems to provide real-time fraud detection and analysis. The service offers pre-built models for text analysis, sentiment analysis, and anomaly detection, and is highly scalable with cloud-based infrastructure. It provides high processing speed suitable for real-time applications and advanced algorithms that ensure high accuracy in detecting anomalies.
  - Spacy can be used to build custom NLP models for specific financial security tasks, such as fraud detection or transaction analysis. The tool provides robust NLP pipelines for tokenization, named entity recognition, and dependency parsing, and allows for extensive customization to meet specific requirements. While it offers efficient processing, it may require optimization for large-scale real-time applications. Spacy shows strong performance in various NLP tasks with customizable pipelines.
  - Stanford NLP can be used for advanced text processing and analysis in financial security systems. It offers comprehensive NLP tools for pars-

ing, sentiment analysis, and entity recognition. The tool is widely used in academic research and provides in-depth linguistic analysis. While it has robust processing capabilities, it may be slower compared to commercial solutions. Stanford NLP is known for its high accuracy in complex NLP tasks and is widely used in academic research.

6.2 Case Studies: NLP in Financial Security

Natural Language Processing in financial security renders numerous advantages that range from fraud detection to customer sentiment analysis, market surveillance, anti-money laundering and risk assessment. NLP assists in financial security, providing real-time analysis and enhancing fraud detection accuracy. As NLP technology advances, its applications in financial security continue to expand, offering innovative ways to protect assets, ensure compliance, and improve customer trust. The subsection discusses implementing NLP in fraud detection with case studies. Two case studies are presented addressing the imbalanced binary classification model and AI-driven model to implement NLP techniques for financial securities. The case studies are useful for understanding real-world problems, complex issues, better learning, problem-solving, critical thinking and developing best practices.

6.2.1 Implementing NLP in Fraud Prevention Systems

Implementing NLP fraud detection needs multiple issues to be considered for the fraud prevention system.

Data Sources for NLP-Based Financial Fraud Detection

- **Transaction Histories**
- **Description:** Transaction histories provide records of individual or bulk financial transactions over time, capturing details such as timestamps, transaction amounts, locations, merchant information, and payment methods.
- **Importance for Fraud Detection:** NLP models analyze transaction descriptions and patterns to identify deviations from typical spending behaviors. For instance, unusual transaction frequencies or foreign purchases could indicate unauthorized activity.
- **Example Use Case**: By processing transaction logs, NLP models can detect keywords or patterns associated with fraud, such as repeated small transactions from unfamiliar locations.

- **User Profiles**
- **Description:** User profiles aggregate information about an account holder, including demographics, usual spending habits, typical locations, and device preferences.
- Importance for Fraud Detection: Behavioral profiling through NLP allows for detecting anomalies by comparing current transaction patterns with historical data. If a user suddenly initiates transactions in a high-risk region, NLP models may flag it as suspicious.
- **Example Use Case:** NLP can leverage user profile data to analyze the semantic meaning of transaction records in the context of usual behavior, helping to improve the accuracy of anomaly detection.
- **Customer Support Logs and Communications**
- **Description:** Customer interactions, such as email correspondence, chat transcripts, and call center logs, contain valuable insights. Support logs may reveal phishing attempts, account takeover attempts, or customer-reported fraud.
- **Importance for Fraud Detection**: NLP models analyze the tone, language, and keywords in these communications. Negative sentiment or urgent language (e.g., "urgent access," "account hacked") can help flag high-risk situations.
- **Example Use Case:** By analyzing keywords and sentiment in customer support logs, NLP models can identify specific fraud-related language patterns and escalate cases that need immediate action.
- **Payment Gateway and API Logs**
- **Description:** API logs from payment gateways include detailed logs of interaction attempts, payment requests, failed authorizations, and system anomalies.
- **Importance for Fraud Detection:** By monitoring language in API logs, NLP models detect irregular requests that could signal a fraud attempt, such as unusual volume or frequency of failed logins, or requests from unfamiliar IP addresses.
- **Example Use Case:** NLP can analyze log entries for unusual behavior, like attempts to bypass transaction limits or use unauthorized payment methods, triggering real-time alerts for fraud investigation.
- **Merchant Information and Risk Scores**
- **Description**: Merchant data, including reputation scores, geographic information, and fraud history, can be vital for risk assessment.
- **Importance for Fraud Detection:** NLP can combine transaction data with merchant risk scores, using past fraudulent associations to flag potential fraud patterns related to specific merchants.
- **Example Use Case:** NLP-based risk analysis can associate specific keywords or transaction descriptions with high-risk merchants, enhancing fraud models' ability to detect transaction anomalies.

In fraud prevention systems, the NLP implementation involves different techniques that analyze unstructured textual data and identify patterns of fraudulent activities. The main steps of the NLP-based Fraud Prevention system are shown in Fig. 6.2. It starts with the collection of textual data then the data preparation is applied where

Fig. 6.2 Flowchart of NLP in general to detect fraud

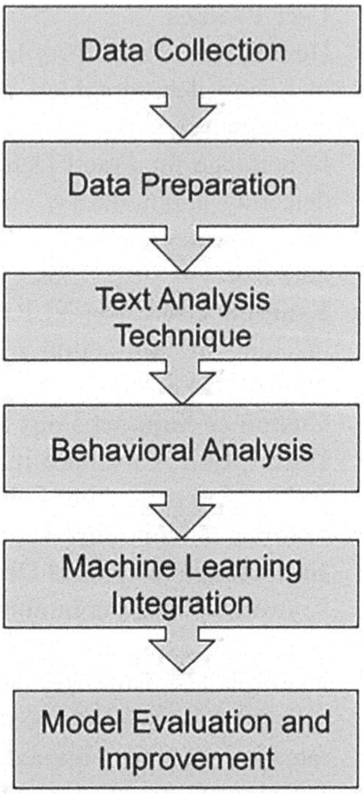

data is preprocessed for consistency and normalization. After the preprocessing of data, different types of textual analysis techniques are applied to process fraud detection. It includes keyword detection, sentiment analysis, anomaly detection, and named Entity recognition. Behavioral analysis is advised to examine sequence patterns and word repetition from users' behavior. The deviation is detected and flagged as fraud. Further machine learning integration is applied, which involves creating learning models to educate the system. The final step is model evaluation for detecting fraud [44].

- STEP 1: Data collection

- The objective of data collection involves gathering textual data that may indicate fraudulent activities. Data collection is a critical phase in implementing NLP for fraud prevention and detection systems because the quality, relevance, and comprehensiveness of the data directly impact the system's effectiveness [45].
- STEP 2: Data Preparation
- The collected data is in raw form and needs cleaning and transformation for suitable NLP analysis. Raw data consists of distortion, inconsistencies and irrelevant information that needs to be addressed. Data preparation consists of several parameters which are shown in the following diagram Fig. 6.3. Data preparation involves filtering out irrelevant information, termed data cleaning, and tokenizing to break down sentences into individual tokens to analyze fraud and signal them. Normalization works on redundancy removal and lemmatization reduces

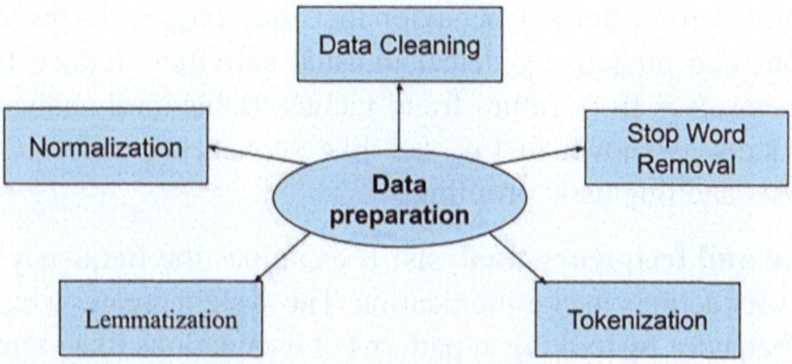

Fig. 6.3 Processes in data preparation

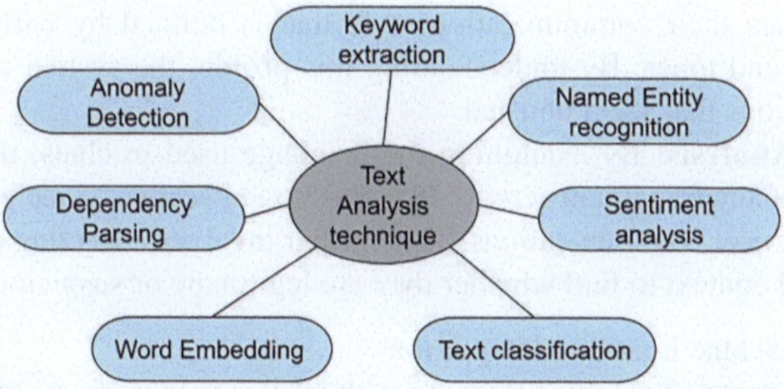

Fig. 6.4 Different ways of text analysis techniques

words to their root form to treat variations of a word similarly like detected and detecting. Additionally, stop word removal assists in reducing noise like "the", "is", "and" which do not carry much significance for fraud detection.

- STEP 3: Text Analysis Technique
- Text Analysis techniques are applied to analyze the text data to find potential fraud indicators by extracting valuable insights from unstructured data. There are different types of text analysis techniques as shown in Fig. 6.4. Identification of specific keywords within the text can be achieved by using keyword extraction techniques. Named entity recognition classifies key entities like name and organization. Word embedding captures semantic linkages and meanings by representing words in a continuous vector space. Anomaly detection detects deviations and classification of the emotional tone in communication is done by sentiment analysis. Grammatical structure analysis of a sentence to determine the relationship between words to understand the meaning of phrases is processed by the dependency parsing technique. In sorting emails into spam or non-spam, predefined categorization or labeling is provided by text classification techniques to detect fraudulent transactions.
- STEP 4: Behavioral Analysis
- User behavior is the most important clue in determining fraud. To detect the anomalies, the behavior is analyzed through their language and patterns. By focusing on the interaction among user communications, organizations can dis-

cover deviations from normal behavior that may require further investigation. Organizations can proactively detect unusual activities, reduce false positives and train themselves from future fraud tactics. Behavioral analysis has several key components as shown in Fig. 6.5 like sequence and frequency analysis, intent analysis and linguistic profiling.

- **Sequence and frequency analysis:** It examines the frequency of keywords, order or user actions in communication. The system creates a baseline of typical user behavior by looking at patterns of user actions like login, transaction, and request to determine how frequently particular phrases occur.
- **Linguistic Analysis:** To identify deviations, this method compares existing communication patterns with the pre-established user profiles. Every user establishes their communication style that is defined by particular words, phrases and tones. By understanding this profile, the system can flag communications that seem unusual.
- **Intent Analysis:** By examining the language used in chats, the techniques may spot any fraudulent activity like phishing or scams, to decipher the actual intent of user communications. This further involves analyzing the emotional tone and context to find whether they are legitimate or suspicious.

- STEP 5: Machine Learning Integration
- The aim of merging machine learning with NLP is to identify fraudulent activity via training models based on linguistic patterns and behavior analysis. The developed system is efficient enough to recognize established and novel frauds [46]. The importance of such integration involves adaptability to evolving threats, scalability, improved accuracy, and real-time detection. Figure 6.6 shows the key components of such integration.
- The main key components in ML integration are learning algorithms and real-time monitoring. Different learning algorithms such as supervised learning, unsupervised learning, semi supervised are applied. Supervised learning classifications are based on labelled datasets. Supervised Learning works on the principle of learning from historical patterns and classifying new transactions. It consists of training of a labelled dataset which is marked either as fraudulent or legitimate. Unsupervised Learning models are particularly useful for finding novice unseen fraud. Semi-supervised learning combines elements of both, utilizing both labelled and unlabeled data in training. It is beneficial when labelled data is limited, as is often the case in fraud detection, where fraudulent cases may

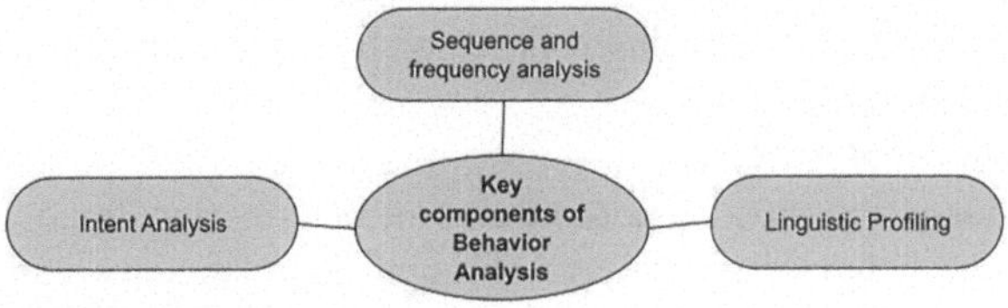

Fig. 6.5 Key components of behavior analysis

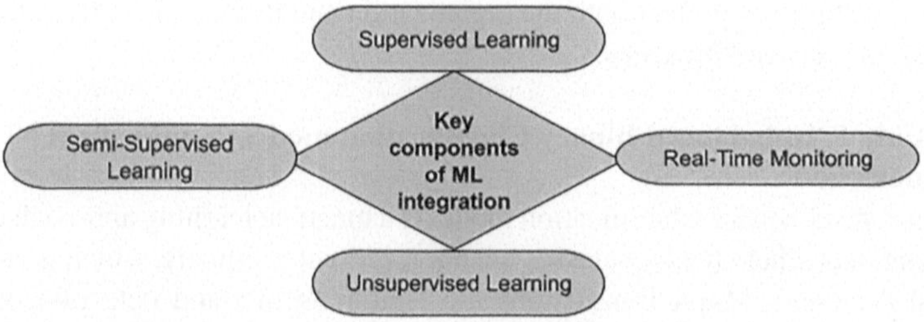

Fig. 6.6 Key components of machine learning integration

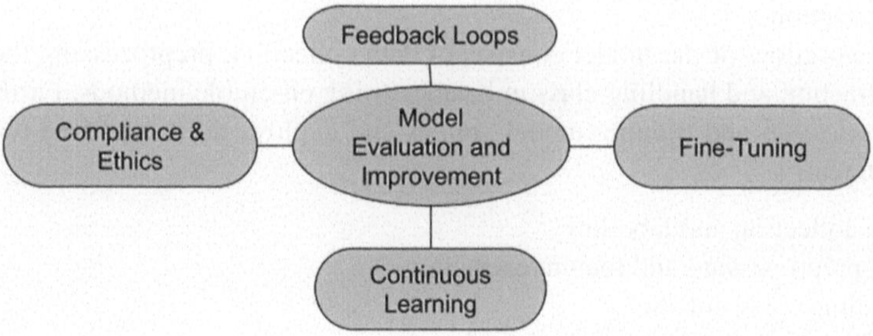

Fig. 6.7 Key components of model evaluation and improvement

be scarce relative to legitimate transactions. Real-time monitoring involves deploying machine learning models in an active environment to analyze transactions as they happen. Organizations can proactively detect fraudulent activity, adjust to new fraud strategies, and improve overall security by utilizing a combination of supervised, unsupervised, and semi-supervised models in addition to real-time monitoring capabilities.

- STEP 6: Model Evaluation and Improvement
- This step ensures that the system is accurate and responsive to emerging fraud patterns [47]. Continuous refinement of the model maintains the effectiveness of the system and makes it adaptable to further changes. It builds trust with users and optimizes resource allocation. It has some key components as shown in Fig. 6.7.
- The main key components of model evaluation and improvement are feedback loops, fine-tuning, compliance & ethics and continuous learning. Under the feedback loop, feedback from real-world transactions allows the system to learn from new fraud cases, adjust model errors and improve accuracy. Fine-tuning detects specific types of fraud like account takeovers, phishing attempts, or unauthorized transactions. Compliance and ethics assure data privacy and transparency regulations. Continuous learning focuses on regular updates and learning from evolving novice data. These models assure accuracy and are adaptable to the latest fraud trends. This ongoing process of assessment and refinement strengthens the

system's ability to protect both the organization and its customers from financial losses and security breaches.

Case Study 1: Imbalanced Binary Classification model to implement NLP in Fraud detection

The imbalanced Binary Classification model is a machine learning approach used to handle datasets where the two classes are not represented equally, which is common in fraud detection. Major transactions are legitimate in fraud detection, while a small percentage are fraudulent. This imbalance can make it challenging for traditional models to learn patterns effectively, as they may become biased towards the majority class (legitimate transactions) and fail to detect the minority class (fraudulent transactions).

The workflow of the model consists of data collection, preprocessing then feature extraction and handling class imbalance using ensemble methods. Further, the model selection and training model applies and deploys the model in a real-time environment.

- Data collection and labelling
- Text preprocessing and feature extraction
- Handling class imbalance
- Model selection and training
- Deploying the Model in a Real-Time Environment
- Continuous Learning and Improvement

This method of detecting fraud in banking views transactions as imbalanced binary classification problems, emphasizing each transaction separately and necessitating a great deal of feature engineering to create user profiles. Adding to this, the procedure extracts logged API calls (such as login and logout) from the bank's server to record user activity. The series of user actions associated with each transaction-related session is documented and categorized according to the bank's classification of the transaction as either genuine or fraudulent. By reducing the requirement for feature engineering and recording the complete activity sequence leading to a transaction, this method turns fraud detection into a sequence classification problem for transfers, and payments. This sequence-based approach makes data preparation easier while also introducing a fresh dataset that contains whole user session sequences. This allows for a more thorough and sophisticated fraud screening process.

Implementing an imbalanced binary classification model in NLP-based fraud detection helps financial institutions accurately detect fraud despite the natural imbalance in transaction data. By leveraging techniques like resampling, cost-sensitive learning, and advanced evaluation metrics, these models can enhance fraud detection capabilities while minimizing disruptions to legitimate transactions. This approach ensures a more efficient, reliable, and adaptive system that protects both the institution and its customers [44].

Case Study 2: AI-Driven Model to Implement NLP in Fraud Detection
The hypothesis of this case study is to protect the E-commerce platforms. In this regard, an AI-driven approach proactively works to identify and mitigate fraudulent activities. The AI-integrated approach provides diverse data collection points and precise analysis of fraudulent behavior. The dataset contains user behavior, purchase history and related variables. Feature engineering methodology is applied to extract relevant data for training AI models using supervised, unsupervised, or semi-supervised learning techniques. Supervised methods like logistic regression classify transactions, while unsupervised approaches detect fraud patterns in unlabeled data. Following training, AI models focus on transactions in real time and flag suspicious acts. Regular model evaluations and updates ensure ongoing accuracy and adaptability, allowing the system to evolve with emerging fraud patterns and maintain its effectiveness. E-commerce platforms are shielded by this approach which guarantees fraud detection and prevention [48]. A defined model is used to carry out the implementation of fraud detection and unauthorized access via examining abnormality [49] in user behavior and analyzing transactional data.

6.2.2 *Lessons Learned from NLP Applications in Financial Security*

Different lessons are learnt from the use of NLP in financial security. The use of NLP plays an active role in detecting and preventing financial fraud. Different NLP-based techniques such as text classification, NER and topic modelling are being used to uncover irregular patterns to identify fraudulent activity. Further, due to the power of context-based semantic analysis, NLP-based models can be used for determining different security issues. Data preprocessing using NLP provides cleaning of data for further accurate analysis. The use of LLM has opened a different path for security in financial documents due to its understanding, comprehension power and learning ability from large and different unstructured data sets. Further use of machine learning has amplified the use of NLP in fraud detection, as supervised, unsupervised, and semi-supervised models analyze and learn from past fraudulent behavior. Real-time monitoring is also possible with different learning-based models which can be trained continuously on real-time data. Different behavior analysis techniques enable fraud detection systems to go beyond surface-level analysis and recognize signs of fraudulent intentions, thereby allowing organizations to act proactively. Different tools based on NLP are also being used.

Finally, to make effective use of NLP in financial documents for security, a balanced data set, continuous training and model evaluation are needed. Fine-tuning models can be applied for phishing or unauthorized transactions. There is a need for compliance and ethical standards enforcement to ensure that fraud detection systems are fair, transparent, and respectful of user privacy. LLMs can also be used for financial security by capturing contextual meanings and aiding in tasks like customer sentiment analysis, market surveillance, and risk assessment.

6.3 Conclusion

As NLP plays a multifaceted role in fraud detection and security in financial documents. This study introduces the use of NLP for fraud detection and security in financial documents. As LLMs are also being applied for security purposes, different aspects and use of LLMs for security concerns in financial documents are also elaborated. Advancement in anomaly and fraud detection using NLP highlighted. Different security measures in NLP are discussed. Different case studies for NLP in financial security were also presented. Finally, a lesson learnt were also presented. The research findings can be applied to prevent and enhance the security of financial documents by using NLP.

References

1. S. S. Mohanrasu, K. Janani, and R. Rakkiyappan, "A COPRAS-based Approach to Multi-Label Feature Selection for Text Classification," *Math Comput Simul*, vol. 222, pp. 3–23, Aug. 2024, https://doi.org/10.1016/j.matcom.2023.07.022.
2. M. Schraagen, M. Brinkhuis, and F. Bex, "Evaluation of Named Entity Recognition in Dutch online criminal complaints," *Computational Linguistics in the Netherlands Journal*, vol. 7, pp. 3–16, 2017.
3. S. Sharma and A. Jain, "Role of sentiment analysis in social media security and analytics," *WIREs Data Mining and Knowledge Discovery*, vol. 10, no. 5, Sep. 2020, https://doi.org/10.1002/widm.1366.
4. J. F. Rodríguez, M. Papale, M. Carminati, and S. Zanero, "A natural language processing approach for financial fraud detection," in *CEUR WORKSHOP PROCEEDINGS*, CEUR-WS. org, 2022, pp. 135–149.
5. L.-C. Chen, C.-L. Hsu, N.-W. Lo, K.-H. Yeh, and P.-H. Lin, "Fraud analysis and detection for real-time messaging communications on social networks," *IEICE Trans Inf Syst*, vol. 100, no. 10, pp. 2267–2274, 2017.
6. F. A. Khan, G. Li, A. N. Khan, Q. W. Khan, M. Hadjouni, and H. Elmannai, "AI-Driven Counter-Terrorism: Enhancing Global Security Through Advanced Predictive Analytics," *IEEE Access*, vol. 11, pp. 135864–135879, 2023, https://doi.org/10.1109/ACCESS.2023.3336811.
7. R. Bridgelall, "Applying unsupervised machine learning to counterterrorism," *J Comput Soc Sci*, vol. 5, no. 2, pp. 1099–1128, Nov. 2022, https://doi.org/10.1007/s42001-022-00164-w.
8. Z. Qiu *et al.*, "Assessing the impact of bag-of-words versus word-to-vector embedding methods and dimension reduction on anomaly detection from log files," *International Journal of Network Management*, vol. 34, no. 1, Jan. 2024, https://doi.org/10.1002/nem.2251.
9. P. Sood, C. Sharma, S. Nijjer, and S. Sakhuja, "Review the role of artificial intelligence in detecting and preventing financial fraud using natural language processing," *International Journal of System Assurance Engineering and Management*, vol. 14, no. 6, pp. 2120–2135, Dec. 2023, https://doi.org/10.1007/s13198-023-02043-7.

10. P. Verma, A. Goyal, and Y. Gigras, "Email phishing: text classification using natural language processing," *Computer Science and Information Technologies*, vol. 1, no. 1, pp. 1–12, May 2020, https://doi.org/10.11591/csit.v1i1.p1-12.
11. S. A. Khushbu, F. T. Jaigirdar, A. Anwar, and O. Tuhin, "Be Aware of Your Text Messages: Fraud Attempts Identification Based on Semantic Sequential Learning for Financial Transactions through Mobile Services in Bangladesh," Jul. 24, 2024. https://doi.org/10.21203/rs.3.rs-4659691/v1.
12. A. Toprak and M. Turan, "Enhanced Named Entity Recognition algorithm for financial document verification," *J Supercomput*, vol. 79, no. 17, pp. 19431–19451, Nov. 2023, https://doi.org/10.1007/s11227-023-05371-4.
13. K. Mishev, A. Gjorgjevikj, I. Vodenska, L. T. Chitkushev, and D. Trajanov, "Evaluation of Sentiment Analysis in Finance: From Lexicons to Transformers," *IEEE Access*, vol. 8, pp. 131662–131682, 2020, https://doi.org/10.1109/ACCESS.2020.3009626.
14. A. Faccia, J. McDonald, and B. George, "NLP Sentiment Analysis and Accounting Transparency: A New Era of Financial Record Keeping," *Computers*, vol. 13, no. 1, p. 5, Dec. 2023, https://doi.org/10.3390/computers13010005.
15. S. Aziz, M. Dowling, H. Hammami, and A. Piepenbrink, "Machine learning in finance: A topic modeling approach," *European Financial Management*, vol. 28, no. 3, pp. 744–770, Jun. 2022, https://doi.org/10.1111/eufm.12326.
16. G. Attigeri, M. P. MM, R. M. Pai, and R. Kulkarni, "Knowledge base ontology building for fraud detection using topic modeling," *Procedia Comput Sci*, vol. 135, pp. 369–376, 2018.
17. J. Su *et al.*, "Large language models for forecasting and anomaly detection: A systematic literature review," *arXiv preprint arXiv:2402.10350*, 2024.
18. W. Xiuguo and D. Shengyong, "An Analysis on Financial Statement Fraud Detection for Chinese Listed Companies Using Deep Learning," *IEEE Access*, vol. 10, pp. 22516–22532, 2022, https://doi.org/10.1109/ACCESS.2022.3153478.
19. B. M. Tornés, E. Boros, A. Doucet, P. Gomez-Krämer, J.-M. Ogier, and V. P. d'Andecy, "Knowledge-Based Techniques for Document Fraud Detection: A Comprehensive Study," 2023, pp. 17–33. https://doi.org/10.1007/978-3-031-24337-0_2.
20. T. Higuchil, A. Gupta, and C. Dhir, "Multi-Task Learning with Cross Attention for Keyword Spotting," in *2021 IEEE Automatic Speech Recognition and Understanding Workshop (ASRU)*, IEEE, Dec. 2021, pp. 571–578. https://doi.org/10.1109/ASRU51503.2021.9687967.
21. Z. Huang, H. Zheng, C. Li, and C. Che, "Application of Machine Learning-Based K-means Clustering for Financial Fraud Detection," *Academic Journal of Science and Technology*, vol. 10, no. 1, pp. 33–39, Mar. 2024, https://doi.org/10.54097/74414c90.
22. S. Subbagari, "Leveraging Optical Character Recognition Technology for Enhanced Anti-Money Laundering (AML) Compliance," *International Journal of Computer Science and Engineering*, vol. 10, no. 5, pp. 1–7, May 2023, https://doi.org/10.14445/23488387/IJCSE-V10I5P102.
23. Z. Lwin Tun and D. Birks, "Supporting crime script analyses of scams with natural language processing," *Crime Sci*, vol. 12, no. 1, p. 1, Feb. 2023, https://doi.org/10.1186/s40163-022-00177-w.
24. Y. Li, S. Wang, H. Ding, and H. Chen, "Large Language Models in Finance: A Survey," in *4th ACM International Conference on AI in Finance*, New York, NY, USA: ACM, Nov. 2023, pp. 374–382. https://doi.org/10.1145/3604237.3626869.
25. K. I. Roumeliotis, N. D. Tselikas, and D. K. Nasiopoulos, "Next-Generation Spam Filtering: Comparative Fine-Tuning of LLMs, NLPs, and CNN Models for Email Spam Classification," *Electronics (Basel)*, vol. 13, no. 11, p. 2034, May 2024, https://doi.org/10.3390/electronics13112034.
26. S. Singhal, A. F. Zambrano, M. Pankiewicz, X. Liu, C. Porter, and R. S. Baker, "De-Identifying Student Personally Identifying Information with GPT-4," in *Proceedings of the 17th International Conference on Educational Data Mining*, 2024, pp. 559–565.

27. A.-R. Akram and D. P. Jyoti, "Revolutionizing Authentication: Harnessing Natural Language Understanding for Dynamic Password Generation and Verification," in *Proceedings of the 20th International Conference on Natural Language Processing (ICON)*, 2023, pp. 670–678.
28. S. Yi *et al.*, "Jailbreak attacks and defenses against large language models: A survey," *arXiv preprint arXiv:2407.04295*, 2024.
29. G. Gambarelli, A. Gangemi, and R. Tripodi, "Is Your Model Sensitive? SPEDAC: A New Resource for the Automatic Classification of Sensitive Personal Data," *IEEE Access*, vol. 11, pp. 10864–10880, 2023, https://doi.org/10.1109/ACCESS.2023.3240089.
30. L. Jiang, "Detecting scams using large language models," *arXiv preprint arXiv:2402.03147*, 2024.
31. C. Jeong, "Fine-tuning and utilization methods of domain-specific llms," *arXiv preprint arXiv:2401.02981*, 2024.
32. "Privacy," https://it.uw.edu/guides/privacy/.
33. R. Subramanian, "Have the cake and eat it too: Differential Privacy enables privacy and precise analytics," *J Big Data*, vol. 10, no. 1, p. 117, Jul. 2023, https://doi.org/10.1186/s40537-023-00712-9.
34. K. Oroy and H. Schield, "Adversarial Attacks and Defenses in NLP: Securing Models Against Malicious Inputs," EasyChair, 2024.
35. A. Brito, C. Fetzer, and P. Felber, "Minimizing latency in fault-tolerant distributed stream processing systems," in *2009 29th IEEE International Conference on Distributed Computing Systems*, IEEE, 2009, pp. 173–182.
36. S. Lai, M. Wang, S. Zhao, and G. R. Arce, "Predicting high-frequency stock movement with differential transformer neural network," *Electronics (Basel)*, vol. 12, no. 13, p. 2943, 2023.
37. G. Bai *et al.*, "Beyond efficiency: A systematic survey of resource-efficient large language models," *arXiv preprint arXiv:2401.00625*, 2024.
38. J. Balsa and G. Lopes, "A distributed approach for a robust and evolving NLP system," in *International Conference on Natural Language Processing*, Springer, 2000, pp. 151–161.
39. R. Patil and V. Gudivada, "A Review of Current Trends, Techniques, and Challenges in Large Language Models (LLMs)," *Applied Sciences*, vol. 14, no. 5, p. 2074, Mar. 2024, https://doi.org/10.3390/app14052074.
40. H. Mao, Y. Liu, Y. Jia, and J. Nanduri, "Adaptive fraud detection system using dynamic risk features," *arXiv preprint arXiv:1810.04654*, 2018.
41. D. Weyns, T. Bäck, R. Vidal, X. Yao, and A. N. Belbachir, "The vision of self-evolving computing systems," *Journal of Integrated Design and Process Science*, vol. 26, no. 3–4, pp. 351–367, Oct. 2023, https://doi.org/10.3233/JID-220003.
42. Vivian Ofure Eghaghe, Olajide Soji Osundare, Chikezie Paul-Mikki Ewim, and Ifeanyi Chukwunonso Okeke, "Navigating the ethical and governance challenges of ai deployment in AML practices within the financial industry," *International Journal of Scholarly Research and Reviews*, vol. 5, no. 2, pp. 030–051, Oct. 2024, https://doi.org/10.56781/ijsrr.2024.5.2.0047.
43. "AI in Banking," https://digitaldefynd.com/IQ/ai-in-banking-case-studies/. Accessed: Jan. 29, 2025. [Online]. Available: https://digitaldefynd.com/IQ/ai-in-banking-case-studies/
44. P. Boulieris, J. Pavlopoulos, A. Xenos, and V. Vassalos, "Fraud detection with natural language processing," *Mach Learn*, vol. 113, no. 8, pp. 5087–5108, Aug. 2024, https://doi.org/10.1007/s10994-023-06354-5.
45. W. Marcellino *et al.*, "Natural Language Processing: Security-and Defense-Related Lessons Learned," 2021.
46. P. Rodriguez and I. Costa, "Artificial Intelligence and Machine Learning for Predictive Threat Intelligence in Government Networks," *Advances in Computer Sciences*, vol. 7, no. 1, pp. 1–10, 2024.
47. L. Wang, Y. Cheng, A. Xiang, J. Zhang, and H. Yang, "Application of Natural Language Processing in Financial Risk Detection," *arXiv preprint arXiv:2406.09765*, 2024.

48. B. Girimurugan, V. Kumaresan, S. G. Nair, M. Kuchi, and N. Kholifah, “AI and Machine Learning in E-Commerce Security: Emerging Trends and Practices,” *Strategies for E-Commerce Data Security: Cloud, Blockchain, AI, and Machine Learning*, pp. 29–53, 2024.
49. M. Roshanaei, M. R. Khan, and N. N. Sylvester, “Enhancing cybersecurity through AI and ML: Strategies, challenges, and future directions,” *Journal of Information Security*, vol. 15, no. 3, pp. 320–339, 2024.

Chapter 7
Multilingual and Cross-Linguistic Challenges in NLP

Dipika Jain

Abstract Natural Language Processing (NLP) has achieved remarkable progress in recent years, with models like BERT, GPT, and others pushing the boundaries of language understanding. However, most advancements have been centered on high-resource languages like English, leaving a significant portion of the world's linguistic diversity underserved. The multilingual and cross-linguistic dimensions of NLP present unique challenges that arise from the vast differences in language structures, data availability, and cultural nuances. This chapter delves into these challenges, exploring issues related to syntactic and morphological diversity, data scarcity for low-resource languages, cross-lingual transfer learning, and the difficulties in evaluating Hindi and Bangla multilingual NLP systems. Through an in-depth examination of these challenges, we aim to shed light on the limitations of current technologies and propose future directions for building more inclusive and equitable NLP models. Addressing these challenges is crucial for expanding the reach of NLP technologies to support global linguistic diversity and foster more accurate, culturally aware language technologies.

7.1 Introduction

The rise of Natural Language Processing (NLP) [1], particularly with the rise of deep learning and transformer-based models like BERT [2], GPT [3], and their multilingual counterparts has a profound impact on our ability to interact with machines through human language. However, much of the early progress in NLP was focused on a few dominant languages, particularly English, which benefits from abundant data, resources, and research. As a result, while NLP systems are highly effective in processing English text [4], they often struggle to perform equally well across the

D. Jain (✉)
School of Computer Science and Engineering, Bennett University,
Greater Noida, Uttar Pradesh, India
e-mail: dipika.jain@bennett.edu.in

A. Kumar, S. R. Sangwan (eds.), *Transformative Natural Language Processing*,
https://doi.org/10.1007/978-3-031-88988-2_7

world's diverse linguistic landscape. The global linguistic environment is incredibly diverse, with more than 7000 spoken languages, many of which are vastly different from one another in terms of syntax, grammar, phonology, and semantics. This diversity introduces a range of cross-linguistic challenges when building NLP systems that aim to be multilingual—that is, capable of processing multiple languages. For instance, a system that works well for languages like English or Mandarin may fail in languages such as Amharic, Xhosa, or Quechua, which lack comparable digital resources.

The significance of addressing these challenges goes beyond just technical curiosity. In an increasingly interconnected world, ensuring that NLP technologies can understand and generate text in many languages is crucial for fostering inclusion, bridging the digital divide, and enabling equitable access to technology. This becomes especially important in domains like healthcare, education, and government services, where the ability to communicate in multiple languages can directly affect people's quality of life. Multilingual NLP [5] also plays a key role in Global communication, social media moderation, Machine translation, and cross-cultural understanding. NLP enables communication across languages through machine translation, facilitating global interactions in business, education, and international relations. Cross-linguistic NLP breaks down language barriers, allowing people to access information and services regardless of their native language.

7.1.1 Importance of NLP

The importance of creating NLP systems that work well across languages is also tied to the preservation of linguistic diversity. Many of the world's languages are underrepresented in the digital realm, and their speakers are at risk of being excluded from technological advancements. Addressing these gaps not only ensures better linguistic coverage but also promotes cultural preservation and prevents the marginalization of smaller language communities. By supporting diverse languages, NLP contributes to maintaining the cultural heritage embedded in linguistic diversity. The ability to process language across multiple languages is crucial for a variety of reasons. In a globalized world, multilingual NLP is a key enabler of cross-cultural communication, whether through machine translation or through digital assistants and chatbots that can interact in different languages. These technologies power a vast array of applications, from international business and social media platforms to public services and educational tools.

Multilingual NLP is also a critical factor in promoting digital inclusion. In regions where local or indigenous languages are the primary mode of communication, the absence of language technologies can severely limit access to information and services. Ensuring that speakers of all languages have access to the benefits of NLP is vital for addressing the digital divide. Moreover, multilingual NLP plays an essential role in preserving endangered languages by digitizing and supporting them in the digital world. Without the development of NLP tools for these languages, they

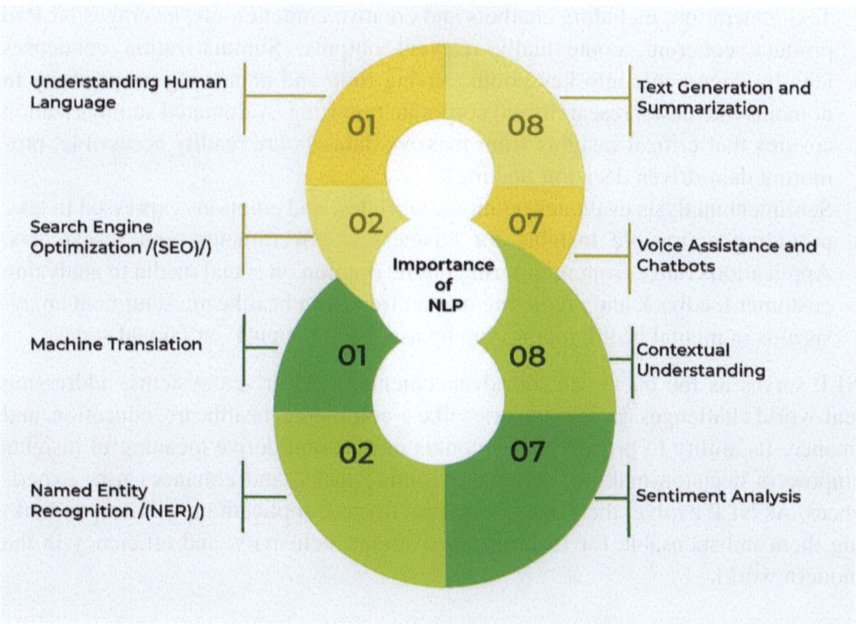

Fig. 7.1 Importance of NLP

risk further marginalization or even extinction. As shown in Fig. 7.1, NLP plays a crucial role in various applications like Machine Translation, Named Entity Recognition, Contextual Understanding, Text generation and summarization, and sentiment analysis etc.

- NLP enables the accurate translation of text between languages, making global communication and accessibility possible. Advanced models like Google Translate or Deep Learning rely on NLP to preserve context, idiomatic expressions, and cultural nuances, facilitating international trade, education, and tourism. Machine translation also helps bridge linguistic gaps in academia, diplomacy, and emergency response by enabling real-time, cross-linguistic communication.
- NER identifies and categorizes entities like names, locations, dates, and organizations in text, aiding in information extraction and data organization. It is critical for applications in legal document analysis, healthcare (e.g., recognizing medical terms), and search engine optimization, as well as for businesses to structure unstructured data and improve decision-making processes.
- NLP powers systems to grasp not only the surface meaning of text but also the underlying context, sarcasm, or implied meaning. This is fundamental for virtual assistants like Alexa and Siri, customer service chatbots, and recommendation systems. Enhanced contextual understanding also supports nuanced applications like legal judgments, clinical diagnosis, and analyzing historical texts for insights.

- Text generation, including chatbots and creative content tools, leverages NLP to produce coherent, contextually relevant outputs. Summarization condenses lengthy documents into key points, saving time and enhancing productivity in domains like news, research, and corporate reporting. Automated summarization ensures that critical insights from massive datasets are readily accessible, promoting data-driven decision-making.
- Sentiment analysis evaluates opinions, attitudes, and emotions expressed in text, providing actionable insights for businesses, governments, and researchers. Applications range from monitoring public opinion on social media to analyzing customer feedback and predicting market trends. In healthcare, sentiment analysis aids in mental health monitoring by assessing patients' emotional states.

NLP serves as the backbone for advancements in AI-driven systems, addressing real-world challenges across industries like e-commerce, healthcare, education, and finance. Its ability to process vast amounts of data and derive meaningful insights empowers decision-making, automates routine tasks, and enhances user experiences. As NLP evolve, their integration into diverse applications will deepen making them indispensable for fostering innovation, inclusivity, and efficiency in the modern world.

7.1.2 Cross-Linguistic Challenges in NLP

Natural Language Processing (NLP) plays a transformative role in cross-linguistics, enabling advanced technical applications in language analysis, communication, and preservation. Machine translation is one of the most prominent uses, where models like Google's Neural Machine Translation (NMT) and Deep Learning employ transformer architectures to achieve contextual accuracy across over 100 languages. Similarly, cross-linguistic sentiment analysis has proven essential for understanding customer opinions globally. Some tools like Sprinklr analyze sentiments from multilingual social media content, leveraging contextual embeddings and multilingual sentiment lexicons to account for cultural nuances. Another critical application is speech recognition, where Automatic Speech Recognition (ASR) models, such as those used in Duolingo and YouTube, transcribe spoken content into text across multiple languages, aiding in education and accessibility.

Cross-linguistic text summarization is gaining traction, particularly in media and research. Platforms like LexisNexis summarize news or research papers from one language to another using abstractive summarization techniques powered by sequence-to-sequence models. Cross-language information retrieval (CLIR) is another essential area, enabling users to query in one language and retrieve relevant documents in another. Google Search and Microsoft's Multilingual Unsupervised and Supervised Embeddings (MUSE) optimize CLIR through multilingual embeddings that align semantic spaces across languages. Meanwhile, addressing low-resource languages remains a significant challenge. NLP also supports linguistic research and endangered language preservation. The Endangered Languages Project

(ELP) and Google's AI for Languages use NLP techniques to digitize and create corpora for rare languages, employing advanced morphological analysis to understand unique linguistic structures. Cross-language question answering (CLQA) is another technical application where models like Facebook AI's XLM-R and datasets such as SQuAD 2.0 facilitate answering questions posed in one language using information available in another. These systems rely on multilingual transformers and attention mechanisms to deliver contextually accurate responses. Multilingual chatbots are similarly transformative in global communication. Platforms like WeChat and Kuki AI deploy conversational agents capable of handling user queries across multiple languages, leveraging bilingual embeddings and transfer learning to maintain conversational coherence.

NLP's capabilities extend to cross-cultural sentiment and emotion analysis, where tools like Alibaba's AI analyze multilingual customer feedback to tailor services according to regional nuances. These systems use sentiment embeddings and emotion lexicons to decode subtle cultural expressions. Despite these advances, cross-linguistic NLP faces challenges such as data scarcity for low-resource languages and cultural diversity in expressions. To address these, pre-trained multilingual models like mBERT, XLM, and LASER have emerged as critical innovations, offering shared semantic spaces across languages.

Cross-linguistic challenges arise from the inherent differences between languages in terms of grammar, syntax, morphology, and semantics. Different languages have different word order rules and grammatical structures. For example, while English follows a Subject-Verb-Object (SVO) pattern, languages like Japanese use a Subject-Object-Verb (SOV) structure, and others like Arabic may employ Verb-Subject-Object (VSO). These variations require NLP systems to adapt to diverse syntactic structures. Languages vary in how they form and inflect words. Languages like Turkish, Finnish, and Hungarian are morphologically rich, meaning that a single word can carry extensive grammatical information. In contrast, English is morphologically poor. NLP systems need to handle this diversity, which complicates tasks like tokenization, parsing, and translation. Words in different languages often carry meanings that are contextually dependent or culturally specific. Translating or processing such words accurately requires an understanding of context, idiomatic expressions, and cultural nuances that go beyond simple dictionary-based translations. NLP powers applications like virtual assistants (e.g., Siri, Google Assistant), chatbots, and customer support systems, enabling them to interact with users in multiple languages. In multilingual contexts, NLP is essential for tools like automatic speech recognition (ASR) [6], sentiment analysis [7, 8], and document summarization.

7.1.3 Data Scarcity and Resource Imbalance

High-resource languages such as English, Mandarin, and Spanish benefit from large corpora, well-developed NLP models, and a wealth of linguistic expertise. On the other hand, many low-resource language, such as Hindi and Bangla, etc. spoken by smaller communities or in specific region mainly suffer from a lack of annotated data, digital text, and linguistic research.

Many languages, including indigenous and regional languages, are considered low resource because they lack large, annotated datasets necessary for training modern NLP models. This scarcity makes it difficult to apply standard machine learning techniques, which typically require large amounts of data to perform well. Even in corpora designed to support multilingual training, the distribution of data is heavily skewed. For instance, large-scale datasets like Common Crawl contain far more data for English and other widely spoken languages than for underrepresented languages. This imbalance results in uneven performance, with NLP systems often performing worse for low-resource languages [5, 9, 10]. For many low-resource languages, the available data may be noisy, incomplete, or poorly structured. This is particularly problematic for languages that have only recently developed written forms, where orthographic conventions may not be standardized. Handling noisy data requires sophisticated pre-processing techniques and careful model design.

7.1.4 Transfer Learning and Multi-Linguistic Models

To address data scarcity, cross-lingual transfer learning has emerged as a promising solution. Transfer learning allows NLP models trained on high-resource languages to be adapted for low-resource languages, leveraging shared linguistic features between them. Multilingual models like mBERT [9, 11], XLM-R [9], and mT5 are designed to process multiple languages, sharing knowledge across languages to improve performance, particularly for underrepresented languages. However, these models face limitations. The performance gap between high-resource and low-resource languages persists, and transfer learning is more effective for languages that are linguistically or typologically similar. In addition to this, the current multilingual models often struggle with the unique challenges posed by languages that differ significantly from those in the training corpus.

We have discussed two of the low-resource languages such as Hindi [5, 9] and Bangla [10] in this chapter as in further sections below.

7.2 Cross-Linguistic Learning Models for Hindi Language

Hindi, being one of the most widely spoken languages in the world, presents a unique set of challenges and opportunities for Natural Language Processing (NLP). As a morphologically rich, inflectional, and free word order language, Hindi is quite different from languages like English, which have been the primary focus of most NLP advancements. Developing models for Hindi requires specialized approaches to handle its syntactic, semantic, and morphological complexities. In this section, we discuss how traditional methods and deep learning models have been applied to Hindi language processing.

7.2.1 *Bag-of-Words and TF-IDF for Text Representation*

In the early days of NLP for Hindi, traditional machine learning models such as Naive Bayes, Support Vector Machines (SVMs), and logistic regression were commonly used for tasks like text classification and sentiment analysis. These models relied on basic text representations, primarily Bag-of-Words (BoW) [12] and Term Frequency-Inverse Document Frequency (TF-IDF) [13] methods.

- **Bag-of-Words:** BoW creates a vocabulary of all words from the training corpus and represents each document as a vector of word frequencies. While BoW is simple and effective for some tasks, it fails to capture the context or semantic relationships between words, which can be critical in a morphologically rich language like Hindi.
- **TF-IDF:** The TF-IDF approach builds on BoW by weighting words based on their frequency across the corpus, helping to downplay common words and emphasize rare or significant terms (Fig. 7.2). While this method is an improvement over BoW, it still treats words as independent entities and struggles with capturing the complexities of Hindi's rich morphology.

Both traditional methods often suffer from limitations when applied to Hindi, particularly in tasks like sentiment analysis or machine translation where the meaning is deeply influenced by context, word inflections, and sentence structure.

	Text	tf	idf
0	Legends speak of a just and brave king.	1	3
1	The lion is often called the king of the jungle	1	3
2	Every king has a rival king to watch out for.	2	3
3	The king met another king at the royal gathering.	2	3

Fig. 7.2 Example for TF-IDF

7.2.2 N-Grams and Feature Engineering

N-gram models is the one which consider sequences of n words to capture local word dependencies were employed in Hindi language processing. However, due to the flexible word order in Hindi, N-grams often face limitations in capturing long-range dependencies and the syntactic structure that is crucial for language understanding.

To mitigate these issues, early machine learning models often relied heavily on feature engineering techniques which involved manually designing features like part-of-speech tags, stemming, lemmatization, and domain-specific lexicons. Feature engineering can improve the performance of model, but it is labour-intensive and specific in terms of language which makes it difficult to generalize to other languages or domains.

7.2.3 Deep Learning Models for Hindi Language Processing

With the advent of deep learning, NLP for Hindi made significant strides. Unlike traditional models, deep learning approaches can automatically learn complex patterns in data without the need for extensive feature engineering. Below are some of the deep learning models used for Hindi language processing.

- **Word Embeddings: Word2Vec, GloVe, and FastText**
- One of the most impactful advancements in NLP has been the development of word embeddings which provide dense, low-dimensional vector representations of words that capture semantic relationships between them. For Hindi, several embedding models have been employed are given as follows:

  (a) *Word2Vec:* This model, based on either Continuous Bag of Words (CBOW) or Skip-gram, generates embeddings by predicting the context of a word (CBOW) or predicting a word given its context (Skip-gram). In Hindi, Word2Vec [14] has been effective in capturing syntactic and semantic relationships but struggles with rare words and out-of-vocabulary terms.
  (b) *GloVe:* Unlike Word2Vec, GloVe [15] (Global Vectors for Word Representation) is a count-based model that uses co-occurrence statistics to generate embeddings. For Hindi, GloVe provides better representations for frequent words but can still suffer from the same limitations with rare or morphologically complex words.
  (c) *FastText:* FastText [16], an extension of Word2Vec developed by Facebook, generates Subwords-level embeddings by incorporating character n-grams.

- **Transformer-Based Models for Hindi**
- In this section, we have discussed few transformer-based models used for hindi language which are as:

(a) *BERT (Bidirectional Encoder Representations from Transformers):* One of the most groundbreaking models in NLP has been BERT, which uses a transformer-based architecture to generate contextual word embeddings. Unlike previous models, BERT captures both left and right context in a sentence simultaneously, allowing it to understand the full context of each word.
(b) *mBERT (Multilingual BERT):* mBERT [9], a version of BERT trained on over 100 languages, including Hindi, has shown great promise in Hindi language tasks. For instance, mBERT performs well in Hindi text classification, named entity recognition (NER), and sentiment analysis due to its ability to capture the contextual meaning of words across different languages.
(c) *Indic BERT:* IndicBERT [9] is a lightweight transformer model specifically designed for Indic languages, including Hindi. It addresses some of the unique challenges posed by the rich morphology and syntactic variations of Indic languages. IndicBERT has shown significant improvements over mBERT in tasks like text classification, NER, and machine translation in Hindi.
(d) *MuRIL (Multilingual Representations for Indian Languages):* Developed by Google, MuRIL is a BERT-based model pre-trained on 17 Indian languages, including Hindi. MuRIL is designed to better handle the complexities of Indian languages, such as code-switching between Hindi and English, a common occurrence in the region.

- **Deep Learning Models for Bangla Language Processing**
- Deep learning models, particularly those based on neural networks, can automatically learn representations of words and phrases without relying on manual feature extraction, making them well-suited for languages like Bangla.

(a) *Word Embeddings: Word2Vec, GloVe, and FastText*

One of the first breakthroughs in NLP was the development of word embeddings, which map words into continuous vector spaces, capturing semantic similarities between words. Several pre-trained word embeddings have been used for Bangla.

- **Word2Vec:** Word2Vec, developed by Google, generates embeddings by predicting a word's context (CBOW) or predicting a word based on its context (Skip-gram). For Bangla, Word2Vec has been useful in tasks like sentiment analysis and text classification, but it struggles with out-of-vocabulary (OOV) words and morphologically complex forms.
- **GloVe:** GloVe (Global Vectors for Word Representation) creates word embeddings by leveraging co-occurrence statistics across large corpora. When applied to Bangla, GloVe captures some word relationships but is less effective than models that handle sub-word information, as it does not consider morphological variations, which are frequent in Bangla.
- **FastText:** FastText, developed by Facebook, extends Word2Vec by considering character-level information. This is particularly useful for Bangla, which has a rich morphology and a highly inflected word system.

FastText embeddings are generated by breaking words into character n-grams, enabling the model to handle unseen words by understanding their sub-word structure. This feature makes FastText more suitable for Bangla, as it can generalize better to new or rare words by using sub-word representations.

(b) *Recurrent Neural Networks (RNNs) and LSTMs*

Recurrent Neural Networks (RNNs) [17] and their variants, particularly Long Short-Term Memory networks (LSTMs), have been applied to many sequence-based tasks in Bangla, such as language modeling, machine translation, and named entity recognition (NER).

- **RNNs and LSTMs**
- RNNs are well-suited for modeling sequential data, such as Bangla sentences, where the meaning of a word often depends on the words preceding or following it. LSTMs, an advanced type of RNN, overcome the problem of vanishing gradients, allowing them to capture long-range dependencies in text. For Bangla language tasks like part-of-speech tagging, NER, and sentiment analysis, LSTMs have been highly effective, as they can learn and retain information over longer sequences, making them better suited to handling the complex syntax and free word order of Bangla. In Bangla-English machine translation, LSTMs have been successful at capturing sentence-level context and handling structural differences between the two languages. However, LSTMs still struggle with very long sequences and complex word inflections that are common in Bangla.
- In cross-lingual sentiment analysis, LSTMs are used to detect sentiment across multiple languages by learning representations that can generalize across linguistic boundaries. For example, an LSTM model can be trained on social media posts, product reviews, or other forms of user-generated content in different languages. These models learn to identify patterns in how sentiment is expressed, even when the words, sentence structures, or cultural context differ. A company could use such a model to monitor brand sentiment across global markets, analyzing feedback in various languages such as English, French, Hindi, or Arabic. This would help the company understand how its products are perceived worldwide, regardless of the language or cultural nuances involved.
- LSTMs are particularly effective in multilingual sentiment analysis because they can manage the sequential nature of text and preserve context across longer passages. When trained on multilingual corpora, LSTMs can detect sentiment in a wide range of languages, even when the specific vocabulary or idiomatic expressions vary. For cross-lingual sentiment analysis, LSTMs often rely on multilingual embeddings—shared vector representations that capture the meanings of words across languages. These embeddings allow LSTM models to transfer knowledge from high-resource languages (such as English or Spanish) to low-resource languages (such as Tamil or Swahili), making it possible to analyze sentiment in languages with limited labelled data. This process, known as cross-lingual transfer learning, enables sentiment analysis to be applied more broadly, without the need for massive annotated datasets in every language.

- Cross-lingual question answering (QA) systems are another area where RNNs and LSTMs have shown significant promise. QA systems are designed to answer user queries by retrieving information from large datasets, often involving vast amounts of text. In a cross-lingual QA scenario, the challenge lies in processing both the questions and answers in different languages. For example, a user might ask a question in English, but the relevant answer could be located in a document written in French, German, or Chinese. LSTMs are particularly useful for handling such tasks because they can process sequential data and understand context over long spans of text. In these systems, LSTM models can be trained to encode a query in one language (such as English) and then retrieve and decode the correct answer in another language (such as French or Chinese). This is typically achieved by leveraging parallel corpora—collections of texts that are available in both source and target languages—or by using cross-lingual embeddings, which enable the model to map both the question and the answer to a shared vector space, regardless of the languages involved. Moreover, in real-world applications like multilingual chatbots and virtual assistants, LSTMs help manage language switching within conversations. Users might switch between languages mid-conversation, and LSTM-based systems can recognize when such switches happen and provide accurate responses in the appropriate language. For example, a multilingual virtual assistant could understand if a user switches from English to Spanish and provide an answer in Spanish without losing the context of the conversation. This capability is critical in a globalized world where people often communicate in more than one language during a single interaction.
- **Bi-directional LSTMs (BiLSTMs)**
- Bidirectional Long Short-Term Memory networks (BiLSTMs) have been employed to Bangla language to effectively capture information from both past and future words in a sequence. This capability is particularly beneficial in tasks such as Named Entity Recognition (NER) and dependency parsing, where understanding the broader context of a word or token is crucial for accurate analysis. Unlike traditional unidirectional models that process text sequentially from left to right (or right to left), BiLSTMs process text in both directions simultaneously, providing a more comprehensive understanding of the surrounding context. For example, in Named Entity Recognition (NER), where the goal is to identify and classify entities like names, locations, dates, or organizations within a sentence, context plays a crucial role. Consider the sentence, "Dhaka is the capital of Bangladesh." A unidirectional LSTM model might miss important contextual cues because it processes words from left to right, treating "Bangladesh" as a potential named entity without realizing that "Dhaka" is the entity it should recognize as the capital. However, with a BiLSTM, the model can simultaneously process the sentence from both directions. By considering the word "Bangladesh" in the future context, the model can better understand that "Dhaka" is a city and not just another common noun. This leads to more accurate recognition of entities and their relationships in Bangla text, which is essential for applications like information extraction, knowledge base creation, and content categorization.

- The application of BiLSTMs in Bangla is in dependency parsing, which involves analyzing the grammatical structure of a sentence and understanding the relationships between words. These systems, which aim to interact with users in Bangla, rely on BiLSTM-based models to accurately interpret user queries and commands. For instance, in a customer service chatbot, users might inquire about product details by mentioning brand names, product specifications, and locations (e.g., "Where can I buy the Apple phone in Dhaka?"). By using BiLSTMs for NER and dependency parsing, the system can effectively identify key entities like "Apple," "phone," and "Dhaka," while also understanding the relationship between them. This ensures the chatbot provides accurate and contextually relevant responses, like store locations, availability, or product features, based on the user's query.
- **Transformer-Based Models for Bangla Language**
- Here, we have discussed about few of the transformer models which is used for processing Bengali language. Transformers excel at capturing long-range dependencies and contextual information, making them ideal for handling the complexities of Bangla.

  (a) *BERT (Bidirectional Encoder Representations from Transformers):* BERT's transformer-based architecture revolutionized NLP by introducing bidirectional training, which allows the model to learn context from both the left and right sides of a token. For Bangla, mBERT (Multilingual BERT) has been employed for a wide range of tasks, including text classification, NER, sentiment analysis, and machine translation.
  (b) *mBERT (Multilingual BERT):* mBERT is pre-trained on over 100 languages, including Bangla, and can be fine-tuned for specific tasks. While it offers significant performance improvements in Bangla tasks, it is often outperformed by models specifically designed for Indian languages or optimized for Bangla (Fig. 7.3). The performance of mBERT transformer model on Bangla tasks is still limited by the scarcity of high-quality Bangla datasets. For many tasks, mBERT requires fine-tuning on Bangla-specific corpora to achieve optimal performance.
  (c) *BanglaBERT:* Recently, Bangla-specific versions of BERT have been developed to address the limitations of multilingual models. BanglaBERT, a pretrained transformer model optimized for Bangla, has shown substantial improvements over mBERT in tasks like Bangla text classification, NER, and question answering. BanglaBERT is trained on a large Bangla corpus, allowing it to better capture the specific syntactic and semantic properties of the Bangla language. It significantly improves performance in downstream NLP tasks compared to general multilingual models, as it is more finely attuned to the linguistic nuances of Bangla.
  (d) *GPT and Language Generation:* OpenAI's GPT-3 [18] and other generative models have also been applied to Bangla, though their use is limited by the availability of large-scale training data in Bangla. GPT-3 can perform tasks like Bangla text generation, dialogue generation, and summarization, though

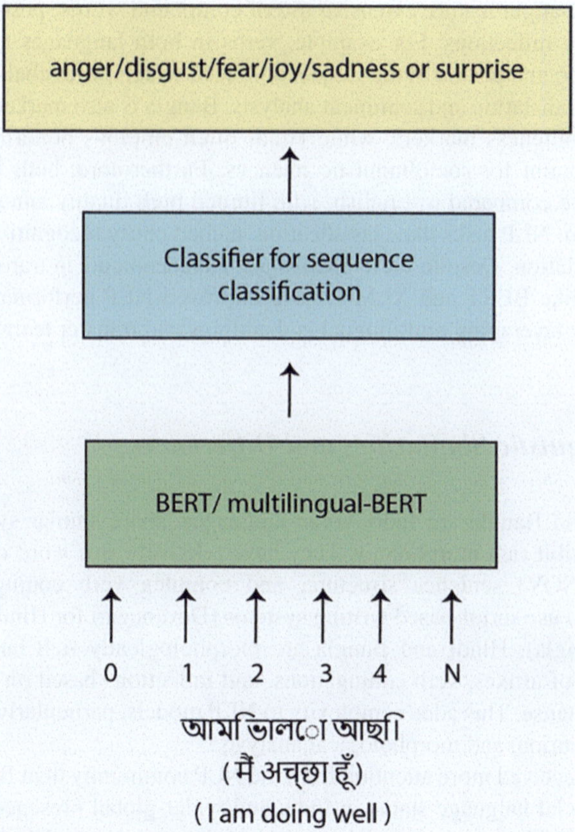

Fig. 7.3 Conceptual architecture of BERT model

its effectiveness is constrained by the amount of Bangla-language data available for fine-tuning.

7.3 Linguistic Challenges and Model Performance for Hindi and Bangla

Natural Language Processing (NLP) for Hindi and Bangla presents unique challenges due to the linguistic intricacies of these two widely spoken South Asian languages. Both languages belong to the Indo-Aryan branch but exhibit significant differences in phonology, syntax, and morphology. Hindi, written in Devanagari script, and Bangla, written in the Bengali script, are morphologically rich and inflectional, making tasks like tokenization, lemmatization, and parsing complex. Their word order follows Subject-Object-Verb (SOV) patterns, but deviations are common due to their flexibility, adding to the syntactic processing difficulties.

Hindi and Bangla feature extensive use of compound words, postpositions, and gender-specific inflections. For example, verbs in both languages agree with the gender, number, and person of the subject or object, which poses challenges in tasks like machine translation and sentiment analysis. Bangla is also marked by its extensive use of politeness markers, while Hindi often employs honorifics, requiring models to account for sociolinguistic nuances. Furthermore, both languages are resource-scarce compared to English, with limited high-quality annotated datasets for downstream NLP tasks like classification, named entity recognition (NER), and machine translation. Despite these challenges, advancements in transformer-based architectures like BERT and XLM-R have improved NLP performance for Hindi and Bangla by leveraging multilingual embeddings and transfer learning.

7.3.1 Linguistic Similarities and Differences

Both Hindi and Bangla are Indo-Aryan languages, share similar syntactic structures, and exhibit rich morphology. They have relatively free word order, subject-object-verb (SOV) sentence structure, and complex verb conjugations. Both languages also use script-based writing systems (Devanagari for Hindi and Bengali script for Bangla). Hindi and Bangla are morphologically rich languages, with extensive use of affixes, verb conjugations, and inflections based on number, gender, case, and tense. This adds complexity to NLP models, particularly for tokenization, lemmatization, and morphological analysis.

Hindi has received more attention from the NLP community than Bangla, largely due to its official language status in India and wider global presence. As a result, more annotated datasets and pre-trained models are available for Hindi. Models like IndicBERT and MuRIL have been developed to specifically handle Hindi and other Indian languages. Bangla is less resourced compared to Hindi, especially in terms of large, annotated corpora, pre-trained language models, and linguistic resources. While models like BanglaBERT have been developed to specifically handle Bangla, they are relatively recent, and research in Bangla NLP is still catching up with Hindi.

7.3.2 Model Performance

In both Hindi and Bangla, traditional ML models like Naive Bayes [19], SVMs [20], and decision trees [21] show moderate success in tasks such as sentiment analysis and text classification. However, they struggle to capture the rich morphological and syntactic properties of these languages. Manual feature engineering, such as using POS tagging and morphological analysers, is often necessary for both languages.

Deep learning models such as Word2Vec, FastText, LSTMs, and transformers like mBERT perform better in both languages due to their ability to learn contextual

Table 7.1 Performance of models for Hindi language

Dataset	Accuracy
Vishesh_Charitr_MBTI [20] (Hindi)	0.668
शब्दसंचित [5]	0.739

relationships and sub-word structures. However, Hindi benefits more from the availability of high-quality pre-trained models like MuRIL and IndicBERT, while Bangla has fewer language-specific models, with BanglaBERT being a recent addition. FastText, with its sub-word modeling, performs well in both languages but is particularly effective for Bangla due to its ability to handle morphologically complex words.

Code-switching between Hindi and English is a common phenomenon, especially in social media and informal texts. Models like MuRIL are designed to handle Hindi-English code-mixed text, a major advantage in this domain. While the code-switching between Bangla and English is also prevalent, especially in Bangladesh and West Bengal. However, Bangla-specific models for handling code-switching are still in their early stages, with less research and fewer resources available compared to Hindi.

Models for Hindi are generally more advanced due to the availability of resources like MuRIL, IndicBERT, and rich annotated datasets. Bangla is lagging but catching up with the development of models like BanglaBERT and the use of FastText for better morphological handling (Table 7.1).

In the above Table 7.1 it compares the predictive accuracy of two datasets for MBTI personality classification in Hindi. The Vishesh_Charitr_MBTI [20] dataset achieves an accuracy of 0.668 (66.8%), reflecting moderate performance, likely influenced by data quality or feature extraction limitations. In contrast, the शब्दसंचित [5] dataset outperforms it with an accuracy of 0.739 (73.9%), indicating a more robust predictive capability. The 7.1% improvement suggests superior data preprocessing, richer linguistic features, or more effective model training. These results underscore the critical role of dataset quality and preparation in enhancing classification accuracy in NLP tasks for low-resource languages like Hindi.

7.4 Other Low-Resource Asian Languages

Asian languages require addressing script diversity, tonal systems, and morphological richness. Advanced multilingual NLP models like mBERT and XLM-R aid in aligning semantic representations across languages. Cross-linguistic NLP applications in these languages not only enhance communication but also preserve linguistic diversity, supporting education, e-governance, and cultural heritage preservation. Owing to their diverse scripts, morphological structures, and linguistic typologies, Asian languages encompass a wide range of linguistic families, including Sino-Tibetan, Dravidian, Turkic, and Austroasiatic, making them an exciting yet complex field for NLP research and application. These languages often pose difficulties related to tokenization, syntax parsing, resource scarcity, and semantic representation, requiring advanced technical solutions such as multilingual embeddings, neural networks, and transfer learning.

One of the most widely studied Asian languages in NLP is Mandarin Chinese, which uses a logographic script with tonal phonology. Machine translation systems like Baidu Translate and Google Translate rely on neural machine translation (NMT) models that incorporate attention mechanisms to handle the contextual meaning of polysemous characters and idiomatic expressions. Additionally, Mandarin's lack of explicit word boundaries creates challenges in segmentation. Japanese introduces complexity with its three scripts—Kanji, Hiragana, and Katakana—combined with its Subject-Object-Verb (SOV) syntax. Morphological analyzers like MeCab and Kuromoji tokenize sentences and resolve ambiguities in polysemous words by leveraging part-of-speech tagging and dictionary-based approaches. Advanced NLP applications in Japanese, such as sentiment analysis, require handling implicit emotions and context-heavy expressions. For instance, tools must account for honorifics and culturally nuanced speech patterns, which significantly affect sentiment scoring. Korean presents challenges related to its agglutinative structure and extensive use of honorifics. Hangul, a phonetic alphabet, simplifies text normalization, but dependency parsing remains complex due to non-linear syntax. Tools like KOMA use graph-based algorithms to handle these parsing tasks effectively. In multimodal NLP, Korean systems integrate ASR with image processing to enhance applications like real-time moderation on social platforms, such as Naver and KakaoTalk. Korean NLP has also advanced in speech synthesis, where neural TTS (Text-to-Speech) models replicate tonal inflections and pitch variations for natural-sounding outputs.

In Southeast Asia, languages such as Thai and Vietnamese highlight unique NLP challenges. Thai, written in an abugida script, lacks spaces between words, making word segmentation critical. PyThaiNLP employs statistical models and deep learning-based approaches to identify word boundaries and improve downstream tasks like Named Entity Recognition (NER). ThaiNER focuses on handling long compound words and cultural references for entity extraction. Vietnamese, on the other hand, uses a Latin-based script with diacritical marks indicating tone. NLP systems must normalize text to account for these diacritics, as tonal changes can alter meanings significantly. ASR systems like Zalo AI and Google Assistant adapt language models to tonal variations, ensuring higher transcription accuracy. Malay and Indonesian (Bahasa), part of the Austronesian family, use a Latin script with relatively simple morphology. However, these languages suffer from data scarcity for high-quality NLP models. Pre-trained transformer models such as IndoBERT have emerged to bridge this gap, employing transfer learning to fine-tune tasks like machine translation, sentiment analysis, and information retrieval. Additionally, e-governance platforms in Southeast Asia use NLP-powered chatbots for multilingual public service, requiring robust cross-lingual understanding.

The Dravidian languages, such as Tamil, Telugu, and Kannada, present unique computational challenges due to their agglutinative nature and extensive inflectional morphology. Written in Brahmic scripts, these languages have complex orthographies that require specialized preprocessing for tasks like tokenization and transliteration. NLP tools such as Google TTS for Tamil focus on preserving phonemic richness while generating natural-sounding synthesized speech. Efforts by AI4Bharat and other initiatives work to improve low-resource datasets, enabling the development of more accurate machine translation and information extraction tools.

In Central Asia, Turkic languages like Uzbek, Kazakh, and Kyrgyz add to the complexity of cross-linguistic NLP due to their use of multiple scripts, including Cyrillic and Latin. NLP models must account for cross-script processing through transliteration and normalization techniques. Tools like Turkic Interlingua focus on creating shared semantic spaces using multilingual embeddings. Low-resource challenges are tackled by leveraging transfer learning and crowdsourced datasets, which improve accessibility for these languages in digital domains. The Tibetan language, spoken in the Sino-Tibetan family, poses difficulties due to its abugida script and limited digital resources. Efforts to develop NLP tools for Tibetan primarily focus on linguistic preservation, using morphological analyzers and annotated corpora to understand its structure. Translation models for Tibetan-English, for instance, rely on hybrid approaches that combine rule-based techniques with neural architectures to overcome the scarcity of parallel data.

Cross-linguistic NLP also supports linguistic research and endangered language preservation. For example, initiatives like the Endangered Languages Project (ELP) digitize and create corpora for rare Asian languages, employing morphological analysis and annotation techniques to preserve their unique structures. These efforts often use unsupervised methods to extract patterns from limited datasets, enabling tasks like text generation and syntactic parsing. A critical challenge across Asian languages is addressing cultural and semantic diversity in cross-linguistic tasks. Multilingual embeddings, such as LASER and XLM-R, align semantic representations across languages by training on large-scale parallel and monolingual corpora. These embeddings allow models to understand idiomatic expressions and cultural nuances in translation and sentiment analysis tasks. Pre-trained models like mBERT and XLM-R play a crucial role in improving cross-lingual understanding, particularly for languages with limited data availability. Despite these advancements, resource scarcity remains a significant obstacle, especially for underrepresented Asian languages. Community-driven initiatives like Masakhane, which focuses on African languages, inspire similar collaborations for Asian NLP. Techniques such as zero-shot learning and transfer learning are increasingly being adopted to extend capabilities to low-resource languages. For instance, multilingual transformers trained on high-resource languages can transfer knowledge to related languages, improving accuracy in tasks like machine translation and question answering.

7.5 Real-Life Customer Services

Cross-linguistic Natural Language Processing (NLP) has emerged as a vital solution for enhancing customer service in India, a country with a rich tapestry of languages and dialects. With over 22 officially recognized languages and hundreds of regional dialects, businesses must find innovative ways to cater to customers who communicate in diverse languages. Cross-linguistic NLP, which enables machines to understand and process multiple languages, has proven to be an essential tool for providing seamless customer service, overcoming linguistic barriers, and improving user experience across India's multilingual landscape. One of the most prominent applications of cross-linguistic NLP in customer service is the development of multilingual chatbots and virtual assistants. These systems are designed to engage with customers in real-time, answering their queries, providing product information, and resolving issues. For instance, telecom companies like Airtel and Jio use chatbots to support a variety of languages, including Hindi, Tamil, Telugu, Bengali, Marathi, and more. The NLP models behind these chatbots, such as mBERT (Multilingual BERT) and XLM-R, can understand the context and intent of a query, even if the user switches languages mid-conversation (a common occurrence known as code-switching). A customer might begin a query in Hindi and switch to English or Tamil midway, and the chatbot will seamlessly continue the conversation, providing accurate and contextually appropriate responses in the correct language. This capability allows companies to offer 24/7 support across multiple languages, ensuring customers have access to help whenever they need it, regardless of their linguistic background.

Cross-linguistic sentiment analysis systems powered by NLP can analyze this feedback in real-time, determining whether it is positive, negative, or neutral. For example, e-commerce platforms like Amazon India or Flipkart can monitor customer reviews in Hindi, Tamil, and English to understand sentiment, identify common complaints, and improve customer experience. Deep learning models, such as BiLSTM (Bidirectional Long Short-Term Memory) or transformers, trained on multilingual datasets, allow businesses to accurately detect sentiment even in code-switched text, such as Hinglish (Hindi and English) ensuring that the underlying emotions of the customer are properly understood. This insight enables companies to act on customer feedback promptly, improving product offerings, resolving issues, and enhancing customer loyalty.

Cross-linguistic NLP also plays a significant role in speech recognition systems used in customer service. Many businesses, especially in industries like banking, telecommunications, and insurance, deploy automated voice systems that handle customer queries through interactive voice response (IVR). Systems like ICICI Bank and HDFC Bank have implemented voice-enabled customer service platforms that recognize and respond to queries in Hindi, Tamil, Bengali, Marathi, Telugu, and more. Cross-linguistic speech recognition systems utilize automatic speech recognition (ASR) models that are capable of transcribing and understanding spoken language in various regional dialects and accents. Whether a customer calls in speaking in Telugu to inquire about their account balance or in Hindi to inquire about a loan, the system processes the voice input, converts it into text, and provides an accurate response. This capability greatly enhances the accessibility of customer service, enabling businesses to serve customers in their preferred language, even in diverse regional contexts.

Another essential application of cross-linguistic NLP in customer service is in email support and ticketing systems. As businesses receive customer inquiries in a variety of languages, cross-linguistic NLP enables systems to automatically detect the language of a customer's query, route it to the appropriate support agent, and generate responses in the same language. For instance, a customer might submit a support request in Kannada regarding an issue with an online purchase, and the system can automatically detect the language, understand the content, and generate an appropriate response in Kannada. This eliminates the need for customers to translate their queries into English or Hindi, making the process smoother and faster. Moreover, multilingual ticketing systems can ensure that customers are not delayed in receiving responses, regardless of the language in which they initiate the request. The use of machine translation systems and cross-lingual embeddings ensures that businesses can provide efficient and accurate customer service in multiple languages, without the need for manual intervention. Personalization in customer service is also greatly enhanced by cross-linguistic NLP. E-commerce companies like MakeMyTrip use cross-linguistic NLP to tailor their customer interactions based on the user's language preferences. By analyzing previous customer interactions, businesses can predict the preferred language of communication and provide personalized recommendations, offers, or product suggestions in that language. For example, if a customer frequently interacts with the platform in Hindi, the system will continue to communicate in Hindi, offering personalized services such as regional hotel deals, travel packages, and promotions tailored to the customer's linguistic and cultural context. This level of personalization improves customer satisfaction and builds stronger relationships with customers by making them feel valued and understood.

7.6 Conclusion

In multilingual and cross-linguistic NLP, English stands out with extensive resources, advanced models like BERT and GPT, and superior performance across tasks. Hindi, though resource-rich compared to other Indian languages, lags English due to its complex morphology and relatively fewer models, such as MuRIL and IndicBERT. Bangla faces even greater challenges, with fewer resources and models like BanglaBERT still in development. While models like FastText help handle Bangla's morphological complexity, both Hindi and Bangla struggle with code-switching and translation tasks. Multilingual models like mBERT offer some solutions, but language-specific models are still crucial for optimal performance. Other low-resource Asian languages, such as Tibetan, Thai, Mandarin, Tamil, Telugu, and Urdu, face similar or even greater challenges in NLP development. Despite their cultural and demographic significance, these languages suffer from limited annotated datasets, underrepresentation in pre-trained multilingual models, and insufficient research attention. For example, Thai's script and tonal nature, Tibetan's complex grammar, and Mandarin's logographic writing system present unique challenges in tokenization and contextual understanding. Tamil, Telugu, and Urdu, with their rich literary histories and morphological diversity, demand specialized models to address nuances in syntax and semantics. Efforts to develop language-specific models like ThaiBERT, TIBERT (for Tibetan), and Indic-focused models are promising but insufficient to fully address the needs of these languages. Advancements in transfer learning, zero-shot learning, and collaborative initiatives to create open-source datasets are crucial for fostering equitable progress in NLP for low-resource languages. The inclusion of these languages in multilingual training frameworks and the design of culturally aware, linguistically robust models will ensure that NLP tools cater to diverse linguistic communities, bridging gaps in accessibility and usability.

References

1. Nadkarni, P. M., Ohno-Machado, L., & Chapman, W. W. (2011). Natural language processing: an introduction. *Journal of the American Medical Informatics Association*, *18*(5), 544-551.
2. Koroteev, M. V. (2021). BERT: a review of applications in natural language processing and understanding. *arXiv preprint arXiv:2103.11943*.
3. Liu, X., Zheng, Y., Du, Z., Ding, M., Qian, Y., Yang, Z., & Tang, J. (2024). GPT understands, too. *AI Open*, *5*, 208-215.
4. Jain, D., Beniwal, R., & Kumar, A. (2024). Advancements in Personality Detection: Unleashing the Power of Transformer-Based Models and Deep Learning with Static Embeddings on English Personality Quotes. *International Journal of All Research Education & Scientific Methods*, *12*(4), 2235-2251.
5. Kumar, A., Jain, D., & Beniwal, R. (2024). HindiPersonalityNet: Personality Detection in Hindi Conversational Data using Deep Learning with Static Embedding. *ACM Transactions on Asian and Low-Resource Language Information Processing*, *23*(8), 1-13.
6. Recognition—ASR, A. S. (2013). Automatic Speech Recognition—ASR.
7. Kumar, A., & Sebastian, T. M. (2012). Sentiment analysis: A perspective on its past, present and future. *International Journal of Intelligent Systems and Applications*, *4*(10), 1-14.
8. Mejova, Y. (2009). Sentiment analysis: An overview. *University of Iowa, Computer Science Department*, 5.
9. Jain, D., & Kumar, A. (2024). AI unveiled personalities: Profiling optimistic and pessimistic attitudes in Hindi dataset using transformer-based models. *Expert Systems*, e13572.
10. Jain, D., Beniwal, R., & Kumar, A. (2023, March). ByaktitbaNet: Deep Neural Network for Personality Detection in Bengali Conversational Data. In *Doctoral Symposium on Computational Intelligence* (pp. 703-713). Singapore: Springer Nature Singapore.
11. Gonen, H., Ravfogel, S., Elazar, Y., & Goldberg, Y. (2020). It's not Greek to mBERT: inducing word-level translations from multilingual BERT. *arXiv preprint arXiv:2010.08275*.
12. Zhang, Y., Jin, R., & Zhou, Z. H. (2010). Understanding bag-of-words model: a statistical framework. *International journal of machine learning and cybernetics*, *1*, 43-52.
13. Yun-tao, Z., Ling, G., & Yong-cheng, W. (2005). An improved TF-IDF approach for text classification. *Journal of Zhejiang University-Science A*, *6*(1), 49-55.
14. Church, K. W. (2017). Word2Vec. *Natural Language Engineering*, *23*(1), 155-162.
15. Pennington, J., Socher, R., & Manning, C. D. (2014, October). Glove: Global vectors for word representation. In *Proceedings of the 2014 conference on empirical methods in natural language processing (EMNLP)* (pp. 1532-1543).

16. Yao, T., Zhai, Z., & Gao, B. (2020, March). Text classification model based on fasttext. In *2020 IEEE International conference on artificial intelligence and information systems (ICAIIS)* (pp. 154-157). IEEE.
17. Sherstinsky, A. (2020). Fundamentals of recurrent neural network (RNN) and long short-term memory (LSTM) network. *Physica D: Nonlinear Phenomena, 404*, 132306.
18. Zhang, M., & Li, J. (2021). A commentary of GPT-3 in MIT Technology Review 2021. *Fundamental Research, 1*(6), 831-833.
19. Back, B. H., & Ha, I. K. (2019). Comparison of sentiment analysis from large Twitter datasets by Naïve Bayes and natural language processing methods. *Journal of information and communication convergence engineering, 17*(4), 239-245.
20. Akshi Kumar, Rohit Beniwal, and Dipika Jain. 2023. Personality Detection using Kernel-based Ensemble Model for Leveraging Social Psychology in Online Networks. ACM Trans. Asian Low-Resour. Lang. Inf. Process. 22, 5, Article 151 (May 2023), 20 pages. https://doi.org/10.1145/3571584
21. Orphanos, G., Kalles, D., Papagelis, A., & Christodoulakis, D. (1999). Decision trees and NLP: A case study in POS tagging. In Proceedings of annual conference on artificial intelligence (ACAI).

Chapter 8
NLP in Action: Case Studies from Healthcare, Finance, and Industry

Shweta and Hifzan Ahmad

Abstract Natural Language Processing (NLP), a branch of artificial intelligence, is transforming various sectors by enabling computers to interpret, analyze, and generate human language. NLP combines linguistics, computer science, and machine learning to support applications like chatbots, sentiment analysis, and text summarization. This review explores NLP's impact across multiple industries, highlighting its ability to improve efficiency, enhance decision-making, and personalize services. In healthcare, NLP aids in patient care, clinical research, and administrative tasks. In finance, it facilitates market analysis, customer service, and fraud detection. The retail sector leverages NLP to analyze consumer behavior and improve customer interactions, while the legal industry benefits from automated document analysis and contract management. Education uses NLP for personalized learning and real-time language assistance, and media relies on it for content creation and audience engagement. In government, NLP improves citizen services, crime prevention, and transparency. While NLP offers numerous benefits, challenges like data privacy and ethical considerations remain areas for further research and development.

Natural language processing (NLP) is a branch of artificial intelligence concerned with how computers and human language interact. It gives machines the ability to comprehend, analyse, and produce human language in a useful manner for applications such as chatbots, sentiment analysis, and translation. Applications such as text analysis,

Shweta (✉)
School of Computer Science and Engineering, Galgotias University, Greater Noida, Uttar Pradesh, India
e-mail: shweta.scse@galgotiasuniversity.edu.in

H. Ahmad
Dr. A.P.J Abdul Kalam Technichal University, Lucknow, Uttar Pradesh, India

A. Kumar, S. R. Sangwan (eds.), *Transformative Natural Language Processing*,
https://doi.org/10.1007/978-3-031-88988-2_8

automated customer service, and speech recognition depend on natural language processing (NLP), which combines linguistics, computer science, and machine learning to enable computers interpret and analyse vast volumes of natural language data.

8.1 Overview of NLP

Natural Language Processing (NLP) is a subset of artificial intelligence that can give computers the ability to read, write, and respond to humans in ways meaningful to both parties. The process breaks down human language into words, phrases, and grammar, making use of several techniques to get the meaning out of the language and determine patterns or even provide answers. NLP techniques start from the rule-based system, the statistical model, all the way to recent and very modern deep learning, particularly with the discovery of the large language model and, more importantly, with the advent of models such as BERT and GPT. Applications of NLP vary from machine translation and sentiment analysis to conversational AI and text summarization and hence are very important in different sectors, including customer services, healthcare, finance, and even social media [1, 2]. However, a lot of problems exist, such as issues regarding context, bias, and less represented languages. Generally, the core research that underlies the development of more sophisticated, multimodal, and ethically robust models capable of forming even more effective relationships with their human partners is being investigated in the future of NLP.

8.2 Healthcare and Medical Industry

NLP significantly transforms the healthcare and medical industry by enhancing patient care, streamlining administrative processes, and improving clinical decision-making [3]. By analyzing unstructured data from medical records, clinical notes, and patient interactions, NLP technologies can extract valuable insights, such as identifying trends in patient symptoms and treatment responses. This capability allows healthcare providers to offer more personalized and effective care plans. NLP also plays a crucial role in automating administrative tasks, such as appointment scheduling and patient inquiries, freeing up healthcare professionals to focus on direct patient care [4–6]. Additionally, NLP aids in clinical research by processing vast amounts of medical literature and extracting relevant information for evidence-based practice. Overall, the integration of NLP in healthcare not only improves efficiency, enhances patient outcomes, and supports better-informed medical decisions.

8.2.1 NLP in Healthcare: Overview

Natural Language Processing (NLP) plays a pivotal role in automating and enhancing clinical workflows, significantly improving the efficiency and effectiveness of healthcare delivery. In the prospective healthcare environment depicted in Fig. 8.1,

Fig. 8.1 Use of NLP at IBM Watson for medical record analysis

NLP technologies improve processes, aid in the analysis of medical records, and offer specific treatment.

- **Automating and Enhancing Clinical Workflows**
- NLP technologies streamline various administrative and clinical processes within healthcare settings. By automating tasks such as data entry, appointment scheduling, and patient triage, NLP reduces the administrative burden on healthcare professionals [7]. This allows them to spend more time on patient care rather than paperwork. Additionally, NLP can facilitate faster retrieval of patient information, enabling healthcare providers to make informed decisions more quickly and effectively [8]. For example, automated transcription services convert doctor-patient conversations into structured electronic health records (EHRs), ensuring accurate and comprehensive documentation without requiring manual input.
- **Natural Language Understanding of Medical Records and Doctor-Patient Interactions**
- NLP excels in the natural language understanding of medical records and doctor-patient interactions. NLP can extract relevant information such as symptoms,

diagnoses, and treatment recommendations by analyzing unstructured data from clinical notes, medical histories, and treatment plans [9, 10]. This capability enhances the accessibility and usability of EHRs, allowing healthcare professionals to identify critical patient information quickly. Furthermore, NLP can analyze doctor-patient interactions to assess sentiment, adherence to treatment plans, and patient engagement, leading to more tailored and responsive care. Overall, NLP's application in understanding medical language fosters improved communication, enhances clinical workflows, and ultimately leads to better patient outcomes.

8.2.2 Case Study 1: Medical Record Automation

IBM Watson has been at the forefront of utilizing NLP to automate the analysis of patient medical records. By processing vast amounts of unstructured data from EHRs, clinical notes, and diagnostic reports, Watson employs advanced NLP algorithms to extract relevant patient information, identify patterns, and generate insights. This technology assists healthcare professionals by synthesizing complex patient histories and presenting the most pertinent data for decision-making [11].

Implementing IBM Watson in medical record automation has enhanced data extraction efficiency and significantly improved diagnostic assistance. Healthcare providers can access crucial patient information more quickly, reducing the time spent sifting through extensive records. This streamlined process not only aids in accurate diagnosis but also allows for more timely treatment interventions. Additionally, the ability to analyze large datasets helps identify potential health risks and recommend personalized treatment plans based on individual patient profiles. Overall, IBM Watson's use of NLP in automating medical record analysis exemplifies how technology can transform healthcare delivery, improve patient outcomes, and optimize clinical workflows. Figure 8.1 represents the automation of medical record analysis using IBM Watson.

8.2.3 Case Study 2: Predictive Healthcare Analytics

NLP technologies are increasingly being employed in predictive healthcare analytics to forecast disease outbreaks and monitor patient deterioration. For instance, systems like HealthMap utilize NLP to analyze data from various sources, including social media, news reports, and public health databases, to detect early signs of disease outbreaks [12]. Similarly, NLP can be applied to analyze electronic health records and patient notes to identify subtle changes in a patient's condition, enabling healthcare providers to anticipate potential health crises before they escalate.

The use of NLP in predictive analytics has resulted in real-time monitoring capabilities that significantly enhance patient care. These systems enable faster intervention and treatment by providing healthcare professionals with timely alerts regarding potential disease outbreaks or indicators of patient deterioration [13–15]. This proactive approach improves patient outcomes by addressing issues before they become critical and enhances public health responses to epidemics by allowing for swift action and resource allocation. Ultimately, the integration of NLP in predictive healthcare analytics fosters a more responsive healthcare environment, leading to better patient management and overall health system efficiency

8.2.4 Key Insights and Future Trends

While the integration of NLP in healthcare presents numerous benefits, it also faces significant challenges. One of the primary concerns is privacy, as healthcare organizations handle sensitive patient information. Ensuring compliance with regulations like HIPAA while leveraging NLP technologies is crucial to maintaining patient confidentiality and trust. Data quality poses a challenge, as NLP systems rely on accurate and comprehensive data to function effectively. Inconsistent or incomplete medical records can lead to incorrect analyses and potentially harmful outcomes, necessitating robust data governance practices to ensure the data's integrity.

Despite these challenges, the future of NLP in healthcare holds considerable opportunities. One promising trend is the advancement of personalized medicine, where NLP can analyze genetic information, patient history, and lifestyle factors to tailor treatment plans to individual patients. This customization level can potentially improve treatment efficacy and reduce adverse effects. Additionally, enhancing patient engagement through NLP-driven tools, such as virtual health assistants and chatbots, can empower patients to take a more active role in their healthcare. These technologies can facilitate better communication between patients and providers, promote adherence to treatment plans, and provide valuable health information. Overall, the ongoing evolution of NLP in healthcare will likely lead to more efficient, effective, and patient-centered care, ultimately improving health outcomes across diverse populations.

8.3 Finance and Banking

NLP is transforming the finance and banking sector by automating processes, enhancing customer service, and improving risk management [16]. Financial institutions leverage NLP to analyze vast amounts of unstructured data, including news articles, social media posts, and customer feedback, to gain insights into market trends and sentiment. Chatbots and virtual assistants powered by NLP provide customers with instant support, helping them with inquiries related to account management, transaction history, and financial advice. Additionally, NLP is used in fraud detection and compliance monitoring by analyzing communication patterns and identifying anomalies that may indicate fraudulent activities [17]. Overall, the integration of NLP in finance not only streamlines operations but also enhances decision-making and customer experiences, positioning financial institutions to respond more effectively to market dynamics.

8.3.1 NLP in Finance: Overview

Natural Language Processing (NLP) plays a critical role in finance by parsing financial documents, automating tasks, and enhancing fraud detection efforts.

- **Parsing Financial Documents**
- NLP technologies enable financial institutions to efficiently analyze and extract information from various types of financial documents, including contracts, reports, and regulatory filings. By utilizing text analysis algorithms, NLP can identify key data points such as terms, conditions, and financial metrics, allowing analysts and decision-makers to access relevant information quickly [18]. This capability significantly reduces the time and effort required for manual document review, improving operational efficiency and enabling faster decision-making.
- **Automating Tasks**
- The automation of routine tasks through NLP is transforming workflows in finance. Chatbots and virtual assistants powered by NLP assist customers with account inquiries, transaction queries, and other banking services, providing real-time support without human intervention. Furthermore, NLP can automate data entry and reporting processes, minimizing errors and freeing up staff to focus on more strategic activities. This streamlining of operations contributes to enhanced productivity and cost savings for financial institutions.
- **Fraud Detection**
- NLP is also instrumental in fraud detection and prevention. By analyzing communication patterns in emails, chat logs, and transaction data, NLP algorithms can identify anomalies and suspicious activities that may indicate fraudulent behavior. This proactive approach allows financial institutions to respond swiftly to potential threats, safeguarding their assets and maintaining customer trust.
- Overall, NLP's applications in parsing documents, automating tasks, and enhancing fraud detection are essential for improving efficiency, accuracy, and security in the finance and banking sector.

8.3.2 Case Study 1: Sentiment Analysis in Stock Markets

In finance, several firms, including MarketPsych and RavenPack, utilize NLP to conduct sentiment analysis on news articles and social media platforms to inform stock market predictions. By employing sophisticated algorithms to process vast amounts of textual data, these systems can evaluate public sentiment regarding specific stocks, sectors, or market events. For example, sentiment derived from tweets, financial news, and online forums can provide insights into investor sentiment and market mood, allowing traders to make more informed decisions based on real-time public opinion.

The integration of NLP-driven sentiment analysis has resulted in improved decision-making within trading algorithms. By incorporating sentiment data into their trading strategies, firms can better anticipate market movements and react quickly to emerging trends or shifts in investor sentiment. This approach has been shown to enhance the accuracy of predictions, enabling traders to capitalize on market opportunities more effectively. Additionally, sentiment analysis allows for the identification of potential risks associated with negative sentiment trends, providing a more comprehensive view of market dynamics. As a result, financial institutions leveraging NLP for sentiment analysis are better positioned to optimize their trading strategies and achieve higher returns on investment.

8.3.3 Case Study 2: Chatbots in Customer Service

Bank of America has implemented an advanced chatbot, Erica, which leverages NLP to enhance customer service interactions. Erica is designed to assist customers with a variety of tasks, including checking account balances, making payments, and providing personalized financial advice. Using NLP, Erica can understand and respond to customer inquiries in natural language, making the interaction more intuitive and user-friendly. The chatbot can also learn from previous interactions to improve its responses over time.

The implementation of Erica has resulted in enhanced customer engagement for Bank of America. Customers appreciate the convenience of having access to immediate assistance without the need to wait for human representatives, thereby improving overall satisfaction. Furthermore, Erica provides 24/7 service, allowing customers to resolve issues and access banking services anytime, regardless of traditional banking hours. This availability increases customer convenience and allows the bank to handle a higher volume of inquiries simultaneously, reducing operational costs. Overall, the use of NLP in chatbots like Erica exemplifies how financial institutions can improve customer experiences and streamline service delivery through innovative technology.

8.3.4 Key Insights and Future Trends in NLP for Finance

While the integration of NLP in the finance sector offers significant benefits, several challenges remain. One major obstacle is the complex financial language used in legal documents, reports, and communications. The nuances and technical jargon inherent in financial texts can complicate NLP systems' accurate parsing and interpretation of information. Additionally, regulatory constraints pose another challenge. Financial institutions must adhere to strict compliance requirements regarding data usage and customer privacy, which can limit the extent to which NLP technologies can be deployed. Balancing the need for innovation with regulatory compliance is crucial for successful implementation.

Despite these challenges, the future of NLP in finance presents numerous opportunities. The rise of robo-advisors is a notable trend, where NLP can be harnessed to provide personalized investment advice based on individual financial situations and goals. These AI-driven platforms can analyze vast amounts of data and generate tailored investment strategies, making financial services more accessible to a broader audience. Furthermore, improvements in fraud detection through NLP technologies are promising. By analyzing communication patterns and transaction data, NLP can more effectively identify anomalies and potential fraudulent activities. As these capabilities continue to evolve, financial institutions can enhance their risk management processes and protect against emerging threats. Overall, embracing NLP technologies will enable finance professionals to improve operational efficiency, enhance customer experiences, and adapt to an ever-changing market landscape.

8.4 Retail and E-commerce

NLP is revolutionizing the retail and e-commerce sectors by enhancing customer experiences, streamlining operations, and optimizing inventory management. Through advanced algorithms, NLP enables businesses to analyze customer feedback, reviews, and social media interactions, providing valuable insights into consumer preferences and trends [19]. Chatbots and virtual assistants powered by NLP facilitate seamless customer service, offering instant support and personalized product recommendations based on user queries and browsing history. Additionally, NLP aids in automating content generation for product descriptions and marketing materials, ensuring consistency and relevance across platforms. By leveraging these capabilities, retailers can improve customer engagement, drive sales, and adapt to changing market dynamics, ultimately creating a more efficient and satisfying shopping experience.

8.4.1 NLP in Retail: Overview

Natural Language Processing (NLP) is a key technology driving innovation in the retail sector by powering recommendation systems, enhancing customer service, and enabling comprehensive review analysis.

- **Recommendation Systems**
- NLP enhances recommendation systems by analyzing customer data, including browsing history, purchase behavior, and textual reviews. By understanding user preferences and detecting patterns in their interactions, NLP algorithms can provide personalized product recommendations. This targeted approach increases customer satisfaction and boosts sales by suggesting relevant items that align with individual tastes and needs.
- **Customer Service**
- In retail, NLP plays a crucial role in improving customer service through chatbots and virtual assistants. These NLP-powered tools can understand and respond to customer inquiries in natural language, providing instant support for a wide range of issues, from product information to order tracking. By automating responses to common questions, businesses can reduce wait times and improve overall customer experience, allowing human agents to focus on more complex inquiries.
- **Review Analysis**
- NLP enables retailers to conduct in-depth customer reviews and feedback analysis across multiple platforms. By employing sentiment analysis, NLP can gauge customer sentiment and identify recurring themes, strengths, and weaknesses in products or services. This valuable information helps retailers make data-driven decisions to enhance product offerings, address customer concerns, and tailor marketing strategies effectively.
- Overall, NLP is transforming the retail landscape by creating more personalized, efficient, and responsive customer experiences, ultimately driving growth and competitiveness in the industry.

8.4.2 Case Study 1: Personalized Product Recommendations

Amazon's recommendation system is a prime example of how NLP can be utilized to enhance personalized shopping experiences. By analyzing vast amounts of customer data, including browsing history, past purchases, and product reviews, Amazon employs NLP algorithms to understand user preferences and behaviors. This system generates tailored product recommendations displayed prominently on the homepage and within individual product pages, allowing customers to discover items they might not have considered.

The implementation of NLP-driven personalized product recommendations has led to a significant **increase in Amazon sales**. By providing customers with relevant suggestions that align with their interests and previous interactions, the platform enhances customer satisfaction and encourages additional purchases. Studies indicate that personalized recommendations can account for a substantial portion of Amazon's overall revenue, demonstrating NLP's effectiveness in creating a more engaging and tailored shopping experience. As a result, this approach has solidified Amazon's position as a leader in the e-commerce industry, showcasing the transformative power of NLP in retail.

8.4.3 Case Study 2: Sentiment Analysis for Consumer Feedback

Retailers like Walmart and Best Buy use NLP-powered sentiment analysis tools to analyze real-time customer reviews and feedback. By processing large volumes of unstructured data from reviews, these tools can detect positive or negative sentiments, uncover specific product issues, and identify consumer preferences. For instance, NLP algorithms can categorize comments about product quality, delivery, or customer service, helping companies better understand what customers like or dislike.

The use of sentiment analysis has led to better product development as retailers can identify common pain points or popular features mentioned in reviews. Armed with these insights, companies can modify existing products or develop new ones that align more closely with customer needs. Additionally, this feedback loop allows retailers to address concerns quickly, leading to improved customer satisfaction. By actively responding to customer sentiment, businesses can enhance their product offerings and foster greater loyalty and trust among their customer base, ultimately driving growth and retention.

8.4.4 Key Insights and Future Trends

While the integration of NLP in retail offers significant benefits, it also presents several challenges. One of the primary concerns is data privacy, as retailers collect and analyze vast amounts of customer data to enhance personalization and recommendations. Striking a balance between leveraging customer insights and maintaining privacy compliance is crucial, especially with regulations like GDPR and CCPA in place. Additionally, providing multilingual support poses a challenge for global retailers. As businesses expand into diverse markets, NLP systems must effectively understand and process multiple languages and cultural contexts to deliver consistent and relevant customer experiences.

Despite these challenges, the future of NLP in retail holds numerous opportunities. One significant trend is the rise of AI-driven personalization, where retailers can utilize advanced NLP algorithms to create highly tailored shopping experiences based on individual customer behavior and preferences. This can lead to increased customer loyalty and higher conversion rates. Moreover, the automation of customer interactions through chatbots and virtual assistants is expected to grow, enabling retailers to provide immediate, round-the-clock support and improve operational efficiency. As NLP technologies continue to evolve, they will likely enable even more sophisticated interactions and insights, further enhancing customer engagement and driving success in the retail sector.

8.5 Legal Sector

NLP increasingly transforms the legal sector by streamlining research, document analysis, and contract management. Through advanced algorithms, NLP can analyze vast amounts of legal documents, case law, and statutes, enabling lawyers to identify relevant information and precedents quickly. Tools that utilize NLP assist in automating tasks such as legal research, contract review, and due diligence by extracting key clauses, summarizing content, and flagging potential issues or risks [20]. Additionally, NLP enhances legal chatbots and virtual assistants, providing clients with quick answers to common legal inquiries and improving access to legal services. Overall, the integration of NLP in the legal field increases efficiency, reduces costs, and allows legal professionals to focus more on strategic decision-making rather than time-consuming administrative tasks.

8.5.1 NLP in Law: Overview

Natural Language Processing (NLP) is playing a crucial role in transforming the legal sector by enhancing document processing, contract analysis, and case law research.

- **Document Processing**

- NLP technologies facilitate the efficient processing of legal documents by automating extracting and organizing relevant information. These tools can analyze unstructured text in contracts, agreements, and legal briefs, identifying key entities, dates, and clauses. This automation streamlines workflows, reduces manual errors, and allows legal professionals to focus on more complex tasks, such as strategy formulation and client engagement.
- **Contract Analysis**
- NLP is particularly valuable in contract analysis, where it assists lawyers in reviewing and managing contracts more effectively. By employing text analysis

algorithms, NLP can automatically identify and highlight critical clauses, inconsistencies, and potential risks within contracts. This capability enables legal teams to conduct thorough reviews in a fraction of the time traditionally required, ensuring compliance and reducing the likelihood of disputes.

- **Case Law Research**
- In legal research, NLP tools enhance the ability to search and analyze vast databases of case law. These technologies can quickly sift through large volumes of legal texts to find relevant cases, statutes, and legal precedents based on specific queries or topics. By providing more accurate search results and summarizing relevant information, NLP allows legal professionals to stay informed of pertinent developments and make informed decisions efficiently.
- Overall, NLP is revolutionizing the legal industry by increasing efficiency, reducing costs, and improving the quality of legal services, ultimately enhancing access to justice.

8.5.2 Case Study 1: Contract Analysis Automation

Kira Systems is a leading provider of contract analysis software that leverages NLP to automate the review of legal documents. By employing machine learning and advanced text analysis techniques, Kira Systems can quickly and accurately identify key contract clauses, terms, and provisions. The platform is designed to help law firms and corporate legal departments streamline their document review processes, enabling legal professionals to upload contracts and receive immediate insights about critical elements, risks, and compliance issues.

The implementation of Kira Systems has led to a significant reduction in time for reviewing contracts, allowing legal teams to process large volumes of documents in a fraction of the time it would typically take. For instance, where a manual review might require several hours or days, Kira's automated analysis can deliver results within minutes, enhancing overall productivity. Additionally, by minimizing human error and ensuring comprehensive analysis, Kira Systems has improved accuracy in contract review, helping legal professionals make better-informed decisions. This automation enhances efficiency and allows legal teams to focus their expertise on more strategic tasks, ultimately leading to improved client satisfaction and more effective legal service delivery.

8.5.3 Case Study 2: Legal Research and Case Prediction

Lex Machina is a pioneering platform that utilizes NLP and machine learning to analyze litigation data and predict legal outcomes. By aggregating and processing vast amounts of historical case data, Lex Machina enables legal professionals to assess trends, judge behavior, and the likelihood of success for specific cases. Users can input various parameters, such as the type of case, jurisdiction, and opposing counsel, and the system provides insights into how similar cases have been resolved, along with potential strategies for the current case.

The use of Lex Machina has led to enhanced case strategy for legal professionals, allowing them to make more informed decisions based on data-driven insights. By understanding the historical context and outcomes of similar cases, lawyers can tailor their arguments, anticipate challenges, and optimize their approach to litigation. Additionally, this predictive capability improves decision-making by equipping legal teams with the knowledge necessary to evaluate the risks and rewards associated with pursuing a particular case. As a result, Lex Machina empowers legal practitioners to formulate more effective strategies, ultimately improving their chances of favorable outcomes in court and enhancing overall client satisfaction.

8.5.4 Key Insights and Future Trends

Despite the promising advancements in NLP within the legal sector, several challenges remain. One significant issue is the ambiguity in legal language, which can complicate the accurate interpretation of contracts, statutes, and case law. Legal terms often have nuanced meanings that may vary by jurisdiction, making it difficult for NLP systems to provide reliable analysis and insights consistently. Additionally, there are ethical concerns surrounding the use of NLP in legal practice, particularly regarding data privacy, bias in decision-making, and the potential for over-reliance on automated systems. Legal professionals must navigate these complexities to ensure that NLP tools are used responsibly and ethically.

On the horizon, there are numerous opportunities for NLP to enhance the legal industry. One of the most significant benefits is the potential for faster and more efficient legal services. By automating routine tasks such as document review and legal research, law firms can reduce operational costs and provide clients with quicker turnaround times. Furthermore, NLP can offer better decision-making tools by providing predictive analytics and insights that inform case strategy, allowing lawyers to make data-driven decisions. As NLP technologies continue to evolve, they will likely lead to greater accessibility in legal services, improved client outcomes, and a more streamlined legal process overall. Embracing these advancements can help legal professionals adapt to the changing landscape and enhance their practice in an increasingly competitive market.

8.6 Education and E-learning

NLP is revolutionizing education and e-learning by enhancing personalized learning experiences, streamlining administrative tasks, and facilitating language learning. Through intelligent tutoring systems, NLP can analyze students' interactions and learning patterns, offering tailored content and feedback that cater to individual needs and learning styles. Additionally, NLP-powered chatbots assist students with inquiries, provide 24/7 support, and automate administrative tasks like grading and scheduling, allowing educators to focus more on teaching. In language education, NLP tools enable real-time translation, pronunciation assistance, and text analysis, helping learners practice and improve their language skills. Overall, NLP is making education more accessible, efficient, and engaging, paving the way for a more dynamic learning environment.

8.6.1 NLP in Education: Overview

Natural Language Processing (NLP) is significantly enhancing the educational landscape by promoting personalized learning, automating grading, and facilitating educational content creation.

- **Personalized Learning**
- NLP enables the development of intelligent tutoring systems that analyze individual student interactions and learning behaviors [21]. These systems can deliver tailored content, resources, and feedback by assessing students' strengths, weaknesses, and preferred learning styles, allowing for a more customized learning experience. This adaptability helps students learn at their own pace and improves overall engagement and retention.
- **Automatic Grading**
- NLP technologies are being utilized to automate the grading of written assignments and assessments. By employing text analysis algorithms, NLP can evaluate student submissions for content accuracy, coherence, and adherence to assignment guidelines. This saves educators significant time and effort and provides students with prompt feedback, enabling them to understand their performance and make necessary improvements.
- **Educational Content Creation**
- NLP tools facilitate educational content creation by generating quizzes, summaries, and interactive learning materials from existing resources. These tools can analyze textbooks, articles, and online resources to produce engaging materials that align with curriculum standards. Additionally, NLP can assist in translating educational content into multiple languages, making learning more accessible to diverse student populations.

- Overall, NLP is transforming education by making learning experiences more personalized, efficient, and engaging, while also streamlining administrative processes and enhancing content accessibility.

8.6.2 Case Study 1: Automated Essay Grading

Grammarly is an exemplary application that utilizes NLP to assist in content evaluation and automated essay grading. By employing advanced algorithms, Grammarly analyzes written text for grammar, punctuation, style, and clarity, providing real-time feedback and suggestions for improvement. Educators can leverage such systems to evaluate student essays and assignments more efficiently. These tools assess various aspects of writing, such as coherence, vocabulary usage, and overall readability, helping educators streamline their grading processes.

The implementation of NLP in automated essay grading systems like Grammarly has led to significant **time savings for educators**, allowing them to focus on more complex aspects of teaching, such as providing personalized feedback and engaging with students. By automating the grading of routine writing assignments, teachers can quickly evaluate large volumes of student work without sacrificing quality. Furthermore, NLP tools offer a more objective evaluation of assignments, minimizing the potential for bias in grading and ensuring that students receive consistent and fair assessments based on predefined criteria. This combination of efficiency and objectivity enhances the educational experience for both educators and students, promoting better learning outcomes.

8.6.3 Case Study 2: Intelligent Tutoring Systems

Duolingo, a popular language-learning platform, employs NLP to enhance its intelligent tutoring system. By analyzing user input and interactions, Duolingo can provide personalized language exercises and assessments tailored to each learner's proficiency level. The platform utilizes NLP algorithms to understand user responses, correct pronunciation, and identify common errors, allowing it to adjust the difficulty of lessons in real time. This adaptive learning approach helps learners progress at their own pace while maintaining engagement.

The integration of NLP in Duolingo has resulted in personalized feedback for users, allowing them to receive instant corrections and suggestions based on their performance. This immediate feedback helps learners understand their mistakes and reinforces learning through practice. Duolingo's intelligent tutoring system also creates adaptive learning paths that evolve as users advance, ensuring that lessons remain relevant and challenging. This tailored approach fosters greater retention and mastery of language skills, making Duolingo an effective tool for language learners worldwide. Overall, NLP enhances the effectiveness of intelligent tutoring systems by providing a more personalized and engaging learning experience.

8.6.4 Key Insights and Future Trends in NLP for Education

One significant challenge in using NLP in education is the potential for bias in educational content. NLP algorithms are trained on existing datasets, which may contain biases or inaccuracies, leading to skewed evaluations and recommendations. This can affect the quality of feedback students receive and perpetuate inequities in educational opportunities. Additionally, language barriers can hinder the effectiveness of NLP tools in diverse classrooms. While NLP can assist in translating and localizing content, nuances in language and context may still pose challenges, making it difficult for non-native speakers to benefit from these technologies fully.

Despite these challenges, the future of NLP in education presents numerous opportunities. One key area is the expansion of adaptive learning technologies that can offer increasingly personalized educational experiences. As NLP capabilities improve, systems can better analyze student data and adapt curricula in real time, creating tailored pathways that cater to individual learning styles and needs. Furthermore, the integration of voice recognition technology can enhance tutoring experiences by allowing for interactive, conversational learning. Students could engage in spoken dialogue with intelligent tutoring systems, receiving immediate feedback on pronunciation and language use, which is particularly beneficial for language learning. As NLP continues to evolve, it will play an essential role in making education more equitable, personalized, and effective.

8.7 Media and Entertainment

NLP is revolutionizing the media and entertainment industry by enhancing content creation, personalization, and audience engagement. Automated tools generate articles, scripts, and subtitles, while recommendation algorithms personalize content based on user preferences [22]. NLP powers voice assistants and chatbots, enabling seamless interaction with entertainment platforms and providing real-time suggestions. It also helps media companies monitor social media, analyze sentiment, and track trends to create more engaging content. Additionally, NLP aids in content moderation, automated translation, and localization, making media more accessible globally. These advancements are transforming how content is produced, distributed, and consumed, offering more tailored and interactive entertainment experiences.

8.7.1 NLP in Media: Overview

Natural Language Processing (NLP) is significantly reshaping the media landscape by enhancing content generation, improving recommendation engines, and enabling sophisticated voice interfaces.

- **Content Generation**
- NLP is utilized in media for automated content creation, where AI-driven tools can generate articles, news summaries, scripts, and even creative writing. These tools analyze existing content to understand styles, tones, and topics, allowing them to produce high-quality drafts that human writers can edit and publish. This speeds up the content production process and helps media organizations keep up with the demand for timely and relevant information.
- **Recommendation Engines**
- NLP plays a critical role in enhancing recommendation engines used by streaming services and media platforms. NLP algorithms can offer personalized content recommendations by analyzing user behavior, preferences, and textual data from reviews and social media interactions. This tailored approach improves user engagement by suggesting movies, shows, and articles that align with individual interests, ultimately increasing viewer satisfaction and retention.
- **Voice Interfaces**
- The integration of NLP into voice interfaces has transformed how audiences interact with media platforms. Voice-activated assistants, such as Amazon Alexa and Google Assistant, utilize NLP to understand natural language commands, allowing users to search for content, play music, or control smart devices using conversational speech. This hands-free interaction enhances user experience, making media consumption more accessible and intuitive.
- Overall, NLP's application in media enhances content quality, personalizes viewer experiences, and simplifies interactions, paving the way for a more dynamic and engaging media landscape.

8.7.2 Case Study 1: Automated Content Creation

The Associated Press (AP) has embraced NLP technologies to automate the creation of news articles, particularly for covering financial reports and sports events. The AP can quickly generate thousands of automated news articles by utilizing an AI-driven platform called Wordsmith. The system analyzes data from earnings reports, game statistics, and other structured information, then produces coherent and informative articles in real-time. This allows the AP to provide timely updates on events that require rapid reporting, such as quarterly earnings or sporting events.

Implementing AI-written articles has led to significantly faster content delivery, enabling the AP to produce articles within minutes of receiving data. This swift turnaround enhances the organization's competitiveness in the fast-paced news environment. Moreover, by automating the writing of routine articles, the AP has reduced the reliance on human journalists for these tasks, allowing them to focus on more complex reporting and in-depth stories. As a result, the AP has increased its overall output while maintaining high journalistic standards, showcasing the effective use of NLP in modern journalism and content creation.

8.7.3 Case Study 2: Speech Recognition for Entertainment

NLP-powered voice assistants like Amazon Alexa and Apple's Siri have transformed how users interact with entertainment platforms. These voice assistants utilize advanced speech recognition and natural language understanding to enable users to control their media consumption through simple voice commands. Users can ask their devices to play specific songs, search for movies, adjust volume levels, and even request personalized recommendations based on their preferences. For instance, saying "Play the latest episode of [show name]" allows for hands-free access to content across various streaming platforms.

Integrating NLP in voice assistants has led to seamless user interactions with entertainment platforms, significantly enhancing the overall user experience. By providing a more intuitive way to navigate content, these voice assistants allow users to access their favorite media effortlessly without needing traditional remote controls or touchscreens. This hands-free approach not only makes media consumption more accessible, especially for those with mobility challenges but also caters to the growing demand for convenient and efficient technology in daily life. As a result, voice assistants have become a key feature in many households, driving engagement with streaming services and reshaping how audiences consume entertainment.

8.7.4 Key Insights and Future Trends

One of the primary challenges in integrating NLP in media is finding the right balance between creativity and automation. While AI-driven tools can efficiently generate content and automate repetitive tasks, there is a risk that the uniqueness and emotional depth inherent in human-created content may be lost. Media organizations must carefully manage the use of NLP technologies to ensure that automated content complements rather than replaces the creative insights and narratives that skilled writers and artists provide. Additionally, maintaining quality standards and authenticity in AI-generated content remains a critical concern, necessitating ongoing oversight and human involvement in the creative process.

The future of NLP in media presents numerous opportunities, particularly in expanding AI-generated content. As NLP models become more sophisticated, they will likely produce higher-quality and more nuanced content across various genres, including journalism, storytelling, and entertainment. This could lead to a significant increase in the volume and diversity of media available to audiences. Furthermore, advancements in NLP can enhance smarter recommendation systems that analyze user preferences and predict emerging trends based on real-time data and audience behavior. This personalized approach to content delivery will foster deeper engagement and satisfaction among users, making it easier for them to discover new and relevant media experiences.

Overall, as NLP technologies continue to evolve, they will play an increasingly vital role in shaping the future of media, balancing the need for efficiency with the demand for creativity and personalized engagement.

8.8 Government and Public Services

NLP is transforming government and public services by improving citizen engagement, automating administrative processes, and enhancing decision-making. Governments provide 24/7 support for public inquiries through chatbots and virtual assistants, while NLP-driven text analysis helps in policy evaluation and public sentiment monitoring [23]. Law enforcement agencies utilize NLP for crime prevention and forensic analysis, while legal and regulatory departments automate document review and ensure compliance. Additionally, NLP is used to process health records, detect fraud in public procurement, and enhance transparency through open data initiatives. However, challenges such as data privacy, security, and mitigating bias in decision-making remain critical considerations for its implementation.

8.8.1 NLP in Government: Overview

NLP is playing a crucial role in transforming how governments operate and deliver public services. By leveraging natural language understanding, governments can automate processes, enhance citizen engagement, and improve decision-making. Here's how NLP aids in three key areas: processing citizen feedback, automating services, and policy analysis.

- **Processing Citizen Feedback**
- NLP allows governments to efficiently manage and analyze large amounts of citizen feedback from surveys, emails, social media, and online forms. Using sentiment analysis and text mining, NLP tools can automatically categorize feedback, detect key issues, and gauge public sentiment. This enables governments to

respond to concerns more rapidly and tailor public services or policies based on real-time input from the population.

- **Automating Services**
- Governments increasingly use NLP to automate routine public services, significantly improving efficiency [24]. NLP-powered chatbots and virtual assistants can handle common citizen inquiries, such as providing information on public services, processing requests like license renewals, or assisting with tax filing. These systems reduce response times, increase service access, and allow human staff to focus on more complex tasks. Additionally, NLP can process large volumes of paperwork and applications by extracting critical information and streamlining administrative procedures.
- **Policy Analysis**
- NLP aids policymakers by analyzing vast amounts of text from legislative documents, research papers, and news sources. It helps extract relevant information, summarize complex reports, and track emerging public opinion and policy discussion trends. Governments use these insights to make more informed, data-driven decisions. NLP tools also help compare policies, identify best practices, and evaluate the potential impacts of legislative changes, thus enhancing the overall policymaking process.
- In summary, NLP enables governments to understand citizen needs better, provide more efficient services, and make informed policy decisions, fostering a more responsive and effective governance system.

8.8.2 Case Study 1: Automating Public Service Requests

Governments worldwide have adopted NLP-powered chatbots to manage routine public service requests. For instance, the Singapore government launched "Ask Jamie," an NLP-based virtual assistant deployed across multiple government agencies. It was designed to handle citizen inquiries regarding public services, such as tax filings, housing applications, and immigration procedures. The chatbot understands and processes natural language queries from citizens, offering instant responses and guiding users through various government services without human intervention.

Implementing NLP chatbots like "Ask Jamie" significantly improved response times, providing instant solutions to common inquiries. This reduced the need for citizens to visit government offices or wait for email responses, improving overall satisfaction with public services. Furthermore, the workload on human customer service agents was significantly reduced, allowing them to focus on more complex or unique cases. In Singapore's case, the chatbot successfully handled over 90% of routine queries, allowing for a more streamlined and efficient public service system. Similar results have been observed in other countries, where NLP automation has helped reduce administrative costs and improve accessibility to services for the public.

8.8.3 Case Study 2: Policy Analysis Using NLP

In the European Union (EU), NLP tools have analyzed extensive legislative documents across multiple languages. One notable example is the use of the European Parliamentary Research Service (EPRS), which leverages NLP to process and summarize large volumes of legislative texts, research reports, and legal frameworks. The tools extract key information, analyze trends, and provide summaries, helping policymakers quickly understand the content of complex legislation and legal amendments.

Using NLP in analyzing legislative texts has resulted in more informed and efficient decision-making. By extracting relevant data and summarizing lengthy documents, policymakers can quickly review essential details without reviewing entire reports. This has led to a more streamlined legislative process, enabling faster responses to emerging political, economic, and social issues. Additionally, NLP tools have made comparing legislative proposals across countries easier, helping policymakers adopt best practices and understand the broader implications of policy decisions. As a result, the EU's ability to craft well-informed and comprehensive policies has improved, directly impacting the quality of governance.

8.8.4 Key Insights and Future Trends

One of the key challenges in using NLP within government settings is ensuring data security. Governments handle sensitive information such as personal data, healthcare records, and financial information, prioritizing data privacy. NLP systems must comply with stringent regulations like GDPR and implement robust encryption to protect citizens' information. Another major challenge is the risk of bias in decision-making. NLP models can unintentionally perpetuate biases present in training data, leading to unfair or discriminatory outcomes, especially in areas like law enforcement, social services, and policy recommendations. Addressing these biases requires transparent algorithms, careful data curation, and continuous monitoring.

NLP presents immense opportunities to enhance public service efficiency. By automating routine tasks such as processing requests, handling inquiries, and managing administrative documents, NLP can significantly reduce the workload on government staff and improve response times. Citizens can access services more quickly and anytime, leading to greater satisfaction and cost savings for governments. Additionally, NLP tools can foster transparency in policymaking by making legislative documents, public records, and data more accessible to citizens. With advanced search and summarization capabilities, NLP can help people better understand complex policies and contribute to more open, accountable governance. The future of NLP in government holds the potential for smarter decision-making, more inclusive services, and a higher degree of public trust.

8.9 Conclusion and Future Outlook

NLP has transformed several industries by making it possible to process language and automate tasks more efficiently. It improves healthcare (diagnostic, patient management), finance (fraud detection, sentiment analysis), education (automatic grading, tutoring), e-commerce (chatbots, recommendations), and cybersecurity (threat detection). While obstacles like data privacy and bias abound, breakthroughs in AI and deep learning continue to increase its applicability. The future of natural language NLP holds the potential of ever more efficient, accurate, and seamless interactions between humans and computers in a variety of sectors.

8.9.1 Cross-Sector Insights

Growth and innovation depend on successfully tackling issues common to all industries.

- Data Quality Challenge: Upholding consistent, correct data is a significant problem in many industries. Inadequate data causes inefficiency and poor decision-making.Solutions: To guarantee data integrity, automate data cleaning procedures, establish robust data governance, and employ real-time monitoring.
- The challenge of privacy and data security: Sensitive data protection and compliance with privacy laws (such as the CCPA and GDPR) are essential issues, as violations can result in severe fines and harm to one's reputation.Solutions: To protect data access, use encryption and anonymization, create robust compliance systems, and use zero-trust security models.
- Technology Integration Challenge: Businesses that depend on conventional IT infrastructure may struggle to integrate new technology with legacy systems.

Insights from various industries underscore the transformative role of NLP in addressing complex challenges and driving innovation. From healthcare to finance and customer service, NLP streamlines operations by automating tasks, enhancing decision-making, and extracting actionable insights from vast amounts of unstructured data. Its impact is reshaping business processes, boosting efficiency, and improving customer experiences.

Looking ahead, NLP is poised for even more significant advancements. As models evolve to understand context and sentiment better, human-AI interactions will become more seamless and natural. Enhanced multilingual support will enable broader global applications, while advanced personalization will offer more tailored and engaging user experiences. Ethical considerations—such as minimizing bias, ensuring transparency, and safeguarding privacy—will be pivotal to its responsible growth. As NLP continues integrating with emerging technologies like AI, IoT, and machine learning, it will unlock new opportunities and efficiencies across various sectors, pushing the boundaries of innovation and transforming old and new industries.

8.9.2 Future Trends in NLP

Advances in making technology more interactive, versatile, and human-like define the changing scope of NLP. There is a growing demand for multimodal NLP, enhancing systems so that they can understand information coming across in varied input types-text, images, audio, and video. This kind of capability makes applications like virtual assistants and interactive media more powerful by understanding both visual and spoken cues. Real-time language translation has another major development: speeding up with the aim of providing more rapid, contextually correct, and culturally relevant translations. This would facilitate global communication in areas such as international business and travel to communicate effectively in multiple languages. Another advancement is conversational agents that can more thoughtfully handle nuanced conversations, show emotional intelligence, and retain memory over several interactions. Virtual assistants and chatbots would then seem much more intuitive and personalized. All of these trends drive NLP towards greater inclusivity, contextual understanding, and real-time responsiveness to form a critical tool both for personal and professional life.

8.9.3 Potential Challenges Ahead

As NLP continues to advance, several challenges are likely to affect its development and integration. Ethical concerns top the list, such as inherent biases in language models that may lead to discriminatory or unbalanced outcomes and job displacement, where automated language technologies increasingly replace certain human roles. Therefore, managing these ethical aspects is crucial to ensuring that NLP serves all users fairly. Technically speaking, it is a huge challenge for technologies because many languages are really complex to deal with; language structures can be highly variable across cultures. Some languages are complicated in their grammatical structure, use many idioms, and carry meaning dependent on contexts, all of which poses a big challenge for models in NLP that need to generalize over such broad linguistic and cultural spaces. Challenges faced: Dealing with these challenges is central to developing NLPs that are fair, effective, and versatile across settings and languages.

8.9.4 *The Road Ahead*

Long-term visions in NLP include promising transformatory potential in such areas as healthcare, finance, education, customer service, and entertainment. The long-term vision in the healthcare domain would involve aiding diagnosis, supporting interaction with patients, and providing accessibility to medical information. Advanced language models can be helpful for automating fraud detection, compliance, and sentiment analysis in finance. NLP-based tools will help improve better personalized learning and provide access to languages in the education sector. Customer service is already seeing the benefits of conversational AI, which will likely be evolved to provide even more sophisticated, emotionally aware support. To achieve these goals, a call to action for ongoing research, innovation, and ethical practices in NLP is crucial, involving challenges such as bias, cultural sensitivity, and the need for inclusive datasets. Through constant responsible development and proactive governance, NLP will then become a powerful universally beneficial technology that respects human communication across industries.

References

1. S. Tian, W. Yang, J. M. L. Grange, P. Wang, W. Huang, and Z. Ye, "Smart healthcare: Making medical care more intelligent," Global Health Journal, vol. 3, no. 3, pp. 62–65, Sep. 2019.
2. T. Young, D. Hazarika, S. Poria, and E. Cambria, "Recent trends in deep learning based natural language processing," arXiv:1708.02709 [cs], Nov. 2018
3. J. Crim, R. McDonald, and F. Pereira, "Automatically annotating documents with normalized gene lists," BMC Bioinformatics, vol. 6, no. 1, p. S13, May 2005.
4. A. Akkasi, E. Varoglu, and N. Dimililer, "ChemTok: A new rule based ˇ tokenizer for chemical named entity recognition," BioMed Research International, vol. 2016, 2016.
5. H. Liu, T. Christiansen, W. A. Baumgartner, and K. Verspoor, "BioLemmatizer: A lemmatization tool for morphological processing of biomedical text," Journal of Biomedical Semantics, vol. 3, p. 3, Apr. 2012.
6. H.-J. Dai, P.-T. Lai, Y.-C. Chang, and R. T.-H. Tsai, "Enhancing of chemical compound and drug name recognition using representative tag scheme and fine-grained tokenization," Journal of Cheminformatics, vol. 7, no. Suppl 1 Text mining for chemistry and the CHEMDNER track, p. S14, 2015.
7. D. Dessi, R. Helaoui, V. Kumar, D. R. Recupero, and D. Riboni, "TFIDF vs word embeddings for morbidity identification in clinical notes: An initial study," arXiv:2105.09632 [cs], Mar. 2020.
8. M. Rahimian, J. L. Warner, S. K. Jain, R. B. Davis, J. A. Zerillo, and R. M. Joyce, "Significant and distinctive n-grams in oncology notes: A text-mining method to analyze the effect of OpenNotes on clinical documentation," JCO Clinical Cancer Informatics, no. 3, pp. 1–9, Dec. 2019.
9. M. Beeksma, S. Verberne, A. van den Bosch, E. Das, I. Hendrickx, and S. Groenewoud, "Predicting life expectancy with a long short-term memory recurrent neural network using electronic medical records," BMC Medical Informatics and Decision Making, vol. 19, no. 1, p. 36, Feb. 2019.
10. M. Hughes, I. Li, S. Kotoulas, and T. Suzumura, "Medical text classification using convolutional neural networks," Studies in Health Technology and Informatics, vol. 235, pp. 246–250, 2017.

11. Somashekhar SP, Sepúlveda MJ, Norden AD et al. Early experience with IBM Watson for Oncology (WFO) cognitive computing system for lung and colorectal cancer treatment. J Clin Oncol 2017;35.
12. Makedon F, Karkaletsis V, Maglogiannis I. Overview: Computational analysis and decision support systems in oncology. Oncol Rep 2006;15:971–974.
13. D. Zhang, J. Thadajarassiri, C. Sen, and E. Rundensteiner, "Timeaware transformer-based network for clinical notes series prediction," in Machine Learning for Healthcare Conference. PMLR, Sep. 2020, pp. 566–588.
14. J. Chu, W. Dong, K. He, H. Duan, and Z. Huang, "Using neural attention networks to detect adverse medical events from electronic health records," Journal of Biomedical Informatics, vol. 87, pp. 118– 130, Nov. 2018.
15. A. Berard, Z. M. Kim, V. Nikoulina, E. L. Park, and M. Gall ´ e, "A ´ multilingual neural machine translation model for biomedical data," in Proceedings of the 1st Workshop on NLP for COVID-19 (Part 2) at EMNLP 2020. Online: Association for Computational Linguistics, Dec. 2020.
16. Tadaaki, H. (2018). Bankruptcy prediction using imaged financial ratios and convolutional neural networks. Expert Systems with Applications, 117, 287–299.
17. Song, Q., Liu, A., & Yang, S. Y. (2017). Stock portfolio selection using learning-to-rank algorithms with news sentiment. Neurocomputing, 264, 20–28.
18. Chong, E., Han, C., & Park, F. C. (2017). Deep learning networks for stock market analysis and prediction: Methodology, data representations, and case studies. Expert Systems with Applications, 83, 187–205
19. J. Hilden, "Statistical diagnosis based on conditional independence does not require it," Computers in Biology and Medicine, vol. 14, no. 4, pp. 429-435, 1984.
20. Hsu, P. Y., Chou, C., Huang, S. H., & Chen, A. P. (2018). A market making quotation strategy based on dual deep learning agents for option pricing and bid-ask spread estimation. The proceeding of IEEE international conference on agents (pp. 99–104).
21. Zhong, H. et al. How does nlp benefit legal system: A summary of legal artificial intelligence. In Proceedings of the 58th Annual Meeting of the Association for Computational Linguistics, 5218–5230 (2020).
22. H. Waguih, "A Data Mining Approach for the Detection of Denial of Service Attack," IAES International Journal of Artificial Intelligence, vol. 2, no. 2, pp. 99-106, 2013
23. Endert, A., Ribarsky, W., Turkay, C., Wong, B. W., Nabney, I., Blanco, I. D., & Rossi, F. (2017, December). The state of the art in integrating machine learning into visual analytics. In Computer Graphics Forum (Vol. 36, No. 8, pp. 458-486).
24. Agerri, R., Artola, X., Beloki, Z., Rigau, G., & Soroa, A. (2015). Big data for Natural Language Processing: A streaming approach. Knowledge-Based Systems, 79, 36-42.

Chapter 9
Generative Large Language Models in Clinical, Legal and Financial Domains

Geetanjali Garg and Shobha Bhatt

Abstract The recent advancements in the field of artificial intelligence (AI) has led to the development of a landmark innovation in the form of generative large language models (LLM). Generative LLMs are the models which take large number of input parameters and generate textual output. Natural language processing (NLP) is a subset of AI that equips machines to comprehend, interpret, and respond to human language. NLP comprises of statistical and deep learning models for providing quality data and understanding the meaning of input. In relation to NLP, generative LLM models can produce contextually correct and relevant response which resembles human-created content given the appropriate prompt. Prompt to these models can be in any form that is unimodal (text, audio or video) or multimodal in nature and will be converted to newer text as output. With the generative power of LLM, NLP can now be used to develop wide range of creative, interactive and dynamic applications ranging from simple translations, conversational tasks to automatic document summarization. The impact of usage of generative LLMs is evident from the advances it has accelerated in high-stakes domain such as healthcare, law and finance. Generative LLMs has stimulated a shift from traditional NLP capabilities to creative content generation capabilities of NLP leading to human like interactions with text. This chapter provides the quantitative understanding of the role of generative LLMs in the field of NLP.

G. Garg (✉)
Department of Software Engineering, Delhi Technological University, New Delhi, Delhi, India
e-mail: geetanjali.garg@dtu.ac.in

S. Bhatt
Department of Computer Science and Engineering, Netaji Subhas University of Technology, East Campus, New Delhi, Delhi, India
e-mail: shobha.bhatt@nsut.ac.in

A. Kumar, S. R. Sangwan (eds.), *Transformative Natural Language Processing*,
https://doi.org/10.1007/978-3-031-88988-2_9

In the recent years, with the tremendous growth in artificial intelligence, the field of human- machine interaction has witnessed major breakthroughs. Machines are required to deal with complex language tasks ranging from information retrieval to translational to real time conversational tasks. For effective human like communication, generalized linguistic models plays an important role. NLP is the branch of AI which conceptualizes human languages through algorithms and models. These models are expected to provide high quality comprehension, interpretation and understanding of natural language to output relevant response. The easy availability of large-scale training data along with the compute intensive NLP based transformer models, has led to the development of Large Language Models that can match the human-level performance and generalizes well on wide range of tasks. LLMs have shown extraordinary capabilities in NLP applications through conversational exchanges with the user. The primary intent of these models is to understand and process natural language for tasks like text translation, classification and retrieval. Bidirectional Encoder Representations from Transformers (BERT) and Text-to-Text Transfer Transformer (T5) are the examples of popular LLMs. These models are capable of comprehending and analyzing text. Further advancements in LLM research resulted in generative version of LLM, popularly known as generative LLM. Generative LLMs has significant role in empowering NLP models by producing complex prompts closely resembling human feedback. In generative LLMs, 'large' indicates the large number of parameters needed to train the models and 'generative' indicates a set of LLMs focussing on generation of text. This generative version of LLM is capable of not only understanding and analysing text but also can generate novel and contextually relevant text output in response to input prompt [1]. Compared to basic LLM like BERT & T5, generative LLM like GPT3 (Generative Pre-trained Transformer) has capability of understanding, processing and generating new text which is often indistinguishable from human generated response by exploring complex patterns within data. NLP and generative LLM both use models from statistics, mathematics and computational linguistics to understand and create what users request. NLP techniques provide quality data to LLMs to avoid misleading user prompts and any kind of controversial content generation. Generative LLMs trained on large corpora spans a wide range of applications, including creation of novel text passages, poetry, question answering, summarization, translations and almost all language related tasks. The research in generative LLMs has shown its vast potential in enhancing approaches used in high-stakes domain such as healthcare, law and finance.

9.1 Exploring Generative LLMs in High-Stakes Domains

The research integrating generative LLMs and NLP has enabled innovative opportunities in different domains. Nowadays, LLMs are being used in high-stakes domains like finance, law and healthcare where consequences of misinterpretation are comparatively high [2]. Domain of financial analytics, medical diagnosis, law, and public safety are the pillars of any social system. These domains are characterised by the high risks, strict compliance to regulations, dependency on professional expertise and experience. The finance domain covers investment strategies, market trends, forecasting, fraud detection, risk management and rigorous financial analysis. Healthcare includes professional expertise, knowledge, understanding, patient management, diagnostics and treatment planning. The legal domain involves understanding of laws, judicial practices, compliance checks, ethical and legal implications. Developing LLMs in all these domains, poses many research challenges relating to accuracy, reliability, interpretability, transparency, data privacy and security, bias and regulatory compliance. Advances in generative LLM research continues to address these challenges while also maintaining transparency to establish trust and reliability in the model outputs. While generative LLMs have the potential of restructuring research methodologies and protocols to provide innovation and efficiency, they also must maintain an adequate balance between innovation and responsibility.

9.1.1 Capabilities and Applications of Generative LLMs

Trained on large corpus of data, generative LLMs exhibit unparalleled capabilities across multiple tasks such as summarising, paraphrasing, question answering, translation and poetry and almost all language related tasks. They output text which resembles human generated content by transforming input prompts into semantically correct and structured output. They can effectively generate novel articles which are relevant in particular context. Since they have the power to interpret user queries, so they can participate in conversation and understand the sentiment behind the written text. They can capture context, identify entities and classify them so they can help in translating and summarizing the documents. The creative content generation capability of LLMs has made them popular choice for writing stories, songs and poetry. Based upon textual descriptions, they are enabled to generate blocks and modules of code and can create synthetic data for machine learning models. Generative LLMs has broadly four major aspects that is language generation, contextual understanding, data augmentation and handling multilingualism which makes them an exciting area to be explored (Fig. 9.1).

9.1.2 Applications of Generative LLM

Generative LLM covers a wide range of applications across various domains including high stakes like healthcare, law, public safety and finance. The versatility in capabilities of LLMs has led humans use them for support in routine decision-making processes. From drafting emails to extracting information from volumes of data, any task involving language processing has been made easier by generative LLMs. These models have shown significant enhancement in various business avenues. Customer service chatbots can understand user queries and generate personalised responses in real time. Businesses can save time and resources by automating generation of product descriptions and market materials at scale. They are also

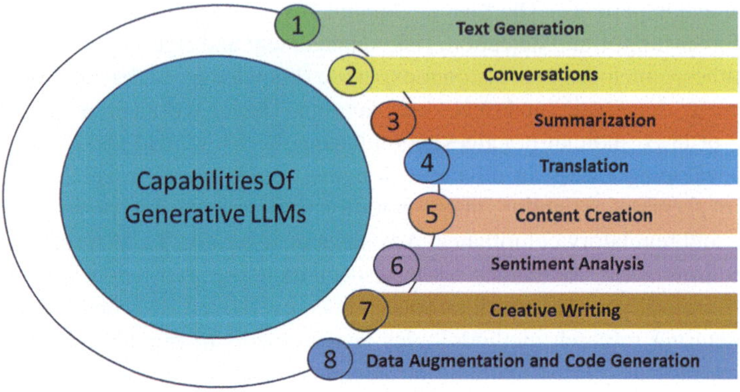

Fig. 9.1 Capabilities of generative LLM

benefitted by automating feedback analysis of customer reviews. The ability of generative LLM to create new representative data can contribute to all research areas where less training data is available. Tasks like summarization of research findings and hypotheses generation can also be performed using generative LLMs. Apart from the usual language processing tasks, generative LLM can produce creative content like poems, stories with different themes. In high stake applications like health care, generative LLMs are being used to automatize the process of clinical notes generation for faster disease diagnosis. Capability of generating hypotheses for new drugs or treatments has open doors for research in the field of drug discovery. In the domain of finance, generative LLMs are being used for generating reports of financial analysis and predicting market trends and future risks. Automatization of drafting work and summarization of legal documents has unburdened legal fraternity from spending hours to complete these jobs. The efficiency with which content is generated by generative LLMs have crossed the limits of what can be produced by an application and accessible to all for specific usage (Fig. 9.2).

9.1.3 Enhancing Content Generation with GPT and Other Models

NLP in its inception, started with extracting information from text assuming it to be a bag of words without focussing on the order or structure of the words in it. To deal with structural component, NLP researchers created word embeddings which are equipped to represent words as numeric vectors but without taking context into consideration. Despite the extensive research on complexities of natural language in NLP, natural language understanding and generation have always remained a challenging area. With the advent of transformer architectures like BERT which are deep neural network designed for NLP tasks, creating contextual word embeddings becomes possible. On the top of BERT transformer, the first generative LLM that is

Fig. 9.2 Application domains of generative LLMs

GPT was developed by Open AI in 2018. Generative LLMs represent a subset of NLP technologies that exploits massive textual data for understanding and generating natural language content.

GPT is a pre-trained transformer model and can be fine-tuned for language modelling and generation task. General transformers comprise of two main modules that is an encoder paired with the input text and a decoder providing output text (Fig. 9.3). GPT has an optimized architecture by having decoder blocks only. Using unsupervised learning techniques, it pre-trains the transformer model on large volume of textual data. For improving model's ability to understand and generate language, a self-attention mechanism is used, allowing the model to consider the context of the entire sentence while generating the next word. The pre-training phase helps the model to learn representations of natural language that can be fine-tuned for specific tasks later.

With the evolution of LLM, Open AI developed a large language model, GPT-3 [3] with the ability of learning distribution of words in text and generate novel text when exposed to large training corpus [4]. In the year 2022, OpenAI launched a ChatGPT [5], a conversational AI system harnessing the power of language processing. Parallel to this, many other effective models like PaLM2 by Google and LLaMA

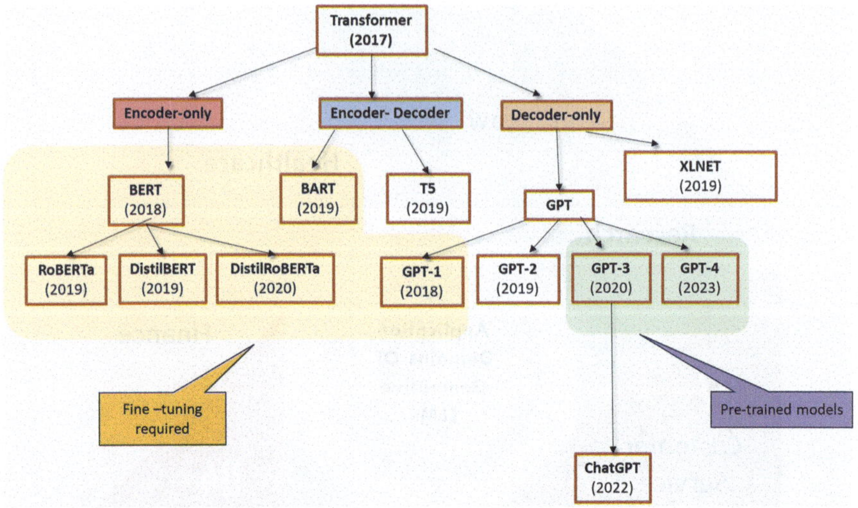

Fig. 9.3 Transformer-based models

by Meta were launched. PaLM2 can handle multilingual data and has improved coding capabilities. Meta's LLaMA is capable of learning through multimodal data. Next, OpenAI released a true innovative solution in the form of GPT-4. GPT4 is multimodal large language model extensively trained on huge number of parameters. Generative LLMs although a new innovative technology but growing at a remarkable speed. The transformative capabilities of generative LLM led to their widespread acceptance for developing applications. Among the primary factors contributing to their growth are the versatility, wide spectrum applicability and scalability. Table 9.1 presents few state-of-the-art generative LLMs.

Although generative LLMs models have great potential of enhancing content creation, they are also associated with certain challenges related to their development process, bias and misuse in future. Overall, GPT and other LLM models covers a range of NLP tasks with high accuracy making them indispensable for various domains including finance, healthcare and many more. As generative AI technology continues to grow at faster pace, the existing language models will improve thereby making human-machine communication more natural and effective.

9.2 Case Studies: Generative LLMs in Action

This section presents case studies in high stake domains that exhibits powerful transformative potential of generative LLMs. Significant advancements using GPT and variants in the fields of healthcare, law and finance have been discussed.

Table 9.1 State of art generative LLM model

Model	Developed by	Features
GPT-4	OpenAI	• Human-like text generation • Wide application ranges from content creation to conversational AI
T5 (Text-to-Text Transfer Transformer)	Google	• Treats NLP problem as a text-to-text problem • Perform multiple tasks such as translation, summarization, and question answering
GatorTronGPT	University of Florida	• Clinical text generation • Trained on a vast corpus of clinical data
BioGPT	Microsoft	• Generating scientific literature summaries for biomedical research
DALL-E	OpenAI	• Generates images from textual descriptions
Codex	OpenAI	• Customised for code generation • Converts natural language prompts into functional code
Llama	Meta	• State-of-the-art open-source large language model • Focussing on responsible usage • Longer context windows • Additional model sizes
Falcon	Technology Innovation Institute of UAE	• Open-source model • Limited multilingual capabilities • Require less memory compared to other models of similar sizes
Mistral Large	Mistral AI	• State-of-the-art reasoning and knowledge capabilities
Gemini	Google	• Multimodal Capabilities • Advanced Reasoning

9.2.1 *Using GPT Models for Clinical Text Generation*

This section presents how generative LLMs are being used effectively for clinical text generation. Healthcare is an area where language plays a crucial role in effective communication between medical professionals and patients. It involves proper documentation of disease history, symptoms, diagnosis, treatment and sometimes life-long care plan. These days with advent of generative LLMs, clinicians may give verbal instructions to computers to write prescriptions, lab tests and create electronic health record (EHR). These systems can also retrieve information from EHR and generate discharge summaries and helps clinicians in faster decision making to provide good and in-time medical advice (Fig. 9.4).

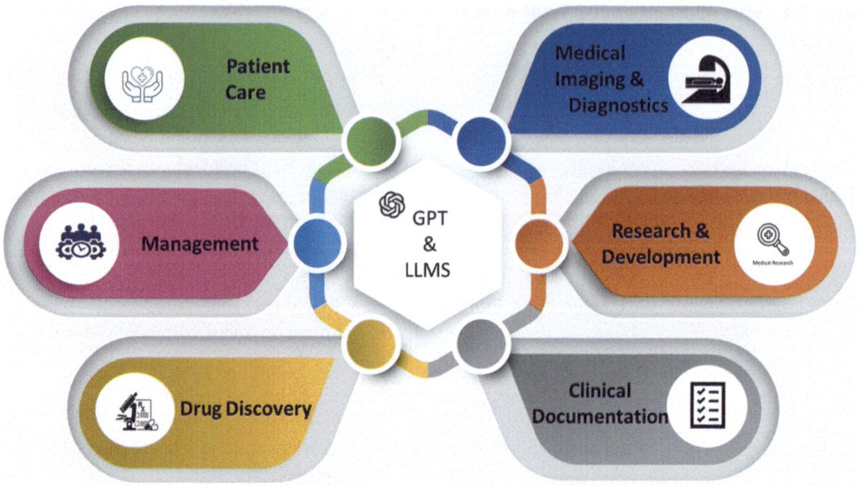

Fig. 9.4 Generative LLMs in healthcare sector

- **Problem statement**
- Clinical text generation is a sub-problem associated with the task of clinical documentation. Clinical documentation involves creation and usage of health records comprising patients' medical history, treatment plan and post-care data. However, the task is generally time-consuming and involves potential risks to patient care. Although the amount of data pertaining to single patient is manually manageable for practitioners, but it becomes a huge burden seeing the increasing number of patients and risks of incorrect entries.
- **Challenges in automating the clinical text generation**
- Automated clinical text generation face challenges in gaining trustworthiness of patients and clinicians, maintaining privacy and safety of data, handling bias in view of life-risks involved with erroneous situations. Clinical text generation has unique challenges as compared to general text generation due to the specific linguistic characteristics i.e. heavy use of professional terminology, abbreviations and acronyms. The lack in good evaluation strategies for the solutions generated by generative LLMs has made the acceptance of these solutions a major problem. Aspects like transparency, performance, and adverse event detection makes these tasks more challenging.
- **Solution as automated clinical text generation**
- Automated clinical text generation require usage of LLMs to create text pertaining to healthcare services. Amalgamating generative LLMs in healthcare industry provide potential benefits to clinicians. Solutions aided by generative LLMs are well researched, faster and less error prone as compared to manual versions. Automated healthcare systems using generative LLMs improves efficiency, accuracy and maintain consistency. The main tasks in automated clinical systems

are sentence classification, question answering, report generation, information extraction, sentence similarity and machine translations. In this direction researchers have explored different LLM models for building clinical domain specific models like GatorTronGPT, ClinGen and ClinicalT5. Chat GPT and its variants are general purpose LLMs and has limitation in addressing domain specific needs. GatorTron GPT [6] was developed by University of Florida and was trained on 82 billion clinical words taken from 126 medical specialities to address issues in healthcare domain. ClinicalT5 [7] is pre-trained text-to-text transformer-based model capable of handling different patterns of writing across clinical documents. Various datasets are developed and being used extensively in training these models. BC5CDR & ChemProt datasets are used for chemical-disease relation and chemical protein extraction. ShARe/CLEF and i2b2-2010 corpus consists of clinical notes for relation extraction. MedSTS dataset is specifically developed for sentence similarity in clinical notes.

- Apart from these, comprehensive evaluation framework like BLUE (Biomedical Language Understanding Evaluation) has been designed to assess the performance of NLP models in the biomedical domain. The BLUE framework can identify diseases, extracting drug to drug interactions, sentence similarity and can also evaluate model's logical reasoning power on clinical text. Few latest research papers also present works in the area of clinical documentation. A user-friendly NLP system tilted Ascle [8] is designed for clinical text generation for biomedical professionals and provide four generative functions comprises of machine translation, question answering, text summarization and text simplification. Ascle system resulted improvement in BLUE score by 20.27 for machine translation task and Rouge-L score by 18% for question answering task. In another study, Dave Van Veen et al. [9], explores the application of LLMs for clinical text summarization specifically for summarizing radiology reports, patient questions and doctor-patient communication. Eight different LLMs were adapted and evaluated for performance against human experts. They also empirically showed the importance of right prompt and its relationship with conciseness of output. The results showed that the best-adapted LLMs fairly performed well in comparison to human experts in terms of correctness of the summaries. They concluded that infusing LLMs into clinical workflows could reduce the documentation burden on clinicians, allowing them to focus more on patient care.
- **Results and benefits of using generative LLM for clinical text generation**

- Although the task of clinical text generation has inherent challenges related to the type of language used, using the power of generative LLM has made it possible to generate realistic human like output. Generative LLMs can generate clinical text at a faster pace making clinicians save upon their time to be spent on documentation task. These systems generate text which is linguistically correct and clinically accurate, consistent with medical regulations and concise. The output generated is also well defined and can serve as a good prompt for future domain specific NLP model. Healthcare professional can focus on patient care and take faster decisions and advice timely treatment plan which is critical aspect in risky situations. Patient management services are getting benefitted from advancements in automated clinical systems in terms of gaining trust, maintaining reliability and privacy of end user.

9.2.2 Legal Document Drafting and Analysis with Generative LLMs

This section presents how generative LLMs are reshaping the legal landscape by making legal processes more efficient and effective. The law firms and professionals can optimize their workflows, enhance productivity, decision making and services. Specifically generative LLMs can aid in drafting legal documents, analysing cases, review contracts, predicting legal outcome, finding relevant precedent and summarization tasks.

- **Problem statement**
- Legal document drafting is the process involving creation and editing of legal documents like agreements, contracts in precise and clear manner while adhering to laws and regulations. Document analysis involves reviewing existing documents to identify implications and potential risks in order to provide good legal advice. Legal documentation and analysis although a routine task but become a major burden as it is associated with volumes of document to be researched and is time-consuming. The professionals need to quickly search references, read, interpret and use them in their ongoing projects and cases for good providing legal advice. Often people engaged in drafting services are non-legal persons making it difficult for them to understand the document and hence prone to errors. Efficiency and time become the major factor for legal professionals to deal with on daily basis in order to retain clients and avoid monetary loss (Fig. 9.5).
- **Challenges in building automated legal solutions**
- Developing an automated legal drafting and analysis system can benefit legal professional to improve upon efficiency and overcome time constraints and provide quality solution to clients on time. Such an automated system comes with many challenges associated with it. Dealing with ambiguity and complex nature of legal language is main challenge for any automated system for legal services. Providing consistent interpretation of legal clause across different versions is another important aspect for legal service automation. Ensuring compliance with laws and regulations and identifying risks is also challenging in legal domain.
- **Solution as automated legal services**
- The use of generative LLMs in legal practices makes it possible to provide more comprehensive legal assistance. These models analyse large number of documents pertaining to previous legal cases and regulations to output accurate and relevant information. By automating drafting and analysis works, error free,

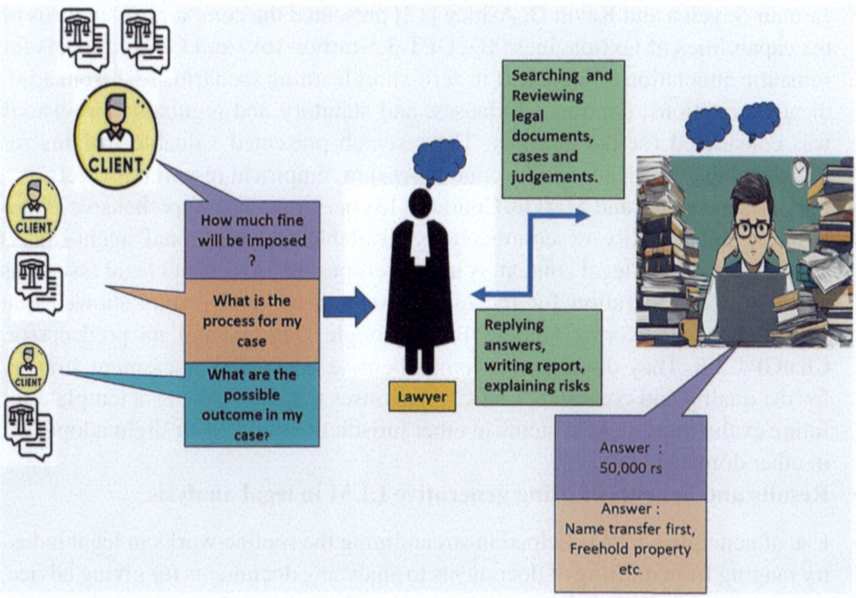

Fig. 9.5 General legal process

high-quality output is obtained in shorter time span thereby allowing professionals to focus on more complex tasks. In this direction researchers have applied GPT and other models in the analysis of legal documents.

- Shabnam Hassani [10] focussed on the usage of LLMs for the analysis of textual legal artifacts. The author conducted an analysis of compliance checking of regulatory artifacts with regulations followed by utilizing LLMs specifically GPT-3.5, GPT-4, Mixtral, and BERT. The content of regulations was taken as textual input for the automated classification of requirements-related legal content and produced corresponding output labels for each provision. A step procedure consisting of preprocessing, LLM-based classification, keyword-based classification, and label prediction was performed. For automated compliance checking, a systematic approach consisting of creation of passages from the regulatory artifact that the user wants to verify for compliance, generation of prompt for LLMs and final step of drawing inferences about compliance and non-compliance was followed. As output a compliance report along with explanation and justification for each determination was generated.
- In another study, Jia-Hong Huang et al. [11] introduced a LLM-based method designed to improve the efficiency of traditional legal proceedings and workflows. The also addressed the problem of scarcity of legal professionals and proposed a new dataset in the domain of legal practices. They conducted empirical analysis using this dataset to show effectiveness of LLM based method. Specially designed prompts for addressing numerical estimation challenges like asset valuation, imprisonment period related to legal decision-making processes were used in this work.

- Jaromir Savelka and Kevin D. Ashley [12] presented the comparative analysis of the capabilities of text-davinci-003, GPT-3.5-turbo(-16k) and GPT-4 models for semantic annotation of legal text in zero-short learning scenario. Text from adjudicatory opinions, contractual clauses, and statutory and regulatory provisions was considered for the analysis. The research presented valuable insights for practical legal applications like contract-review, empirical research projects.
- Sascha Schweitzer and Markus Conrads [13] performed a comprehensive evaluation of applicability of commercially available conversational agents (CA) within the German legal context. A unique corpus of 200 distinct legal tasks was taken into consideration for the qualitative analysis and results showed that ChatGPT-4 outperforms Google Bard, Google Gemini, and its predecessor, ChatGPT-3.5. They developed a comprehensive query and assessment strategy for the quality and consistency of CA responses that can serve as a template for future evaluations of AI systems in other jurisdictions and, with slight adoptions, in other domains.
- **Results and benefits of using generative LLM in legal analysis**

- Use of generative LLMs helped in streamlining the routine works in legal industry ranging from drafting of documents to analysing documents for giving advice. Automation of these tasks has enhanced productivity and accuracy. By implementing LLM driven solutions professionals save upon their time spent on searching legal databases for relevant documents. Automated system can handle repetitive tasks and reduce workload on legal team and thereby enabling them to spend time on tasks which require human expertise. The chances of human error are greatly reduced and hence improve the precision and accuracy of legal documents. Faster solution and low operational cost of automated system has increased the number of clients being served in fixed period leading to monetary benefits to law firms. The Clients have better satisfaction level, and this has helped in retention of clients for a longer period. Lawyers can make use of comprehensive data analysis provided by generative LLM powered solutions. Huge volume of documents can be analysed faster and enables legal professionals to take better decisions. The prompt response with accuracy has enabled legal industry to improve their overall business performance. The Key benefits from automated legal system are

  - Reduced time
  - Increased Efficiency
  - Enhanced Accuracy
  - Minimized revenue loss
  - Minimise operational cost
  - Increased client retention
  - Scalable services
  - Better decision making

  - Generative LLMs can be used in other applications apart from document drafting and analysis in legal sector (Fig. 9.6). Major being the contract

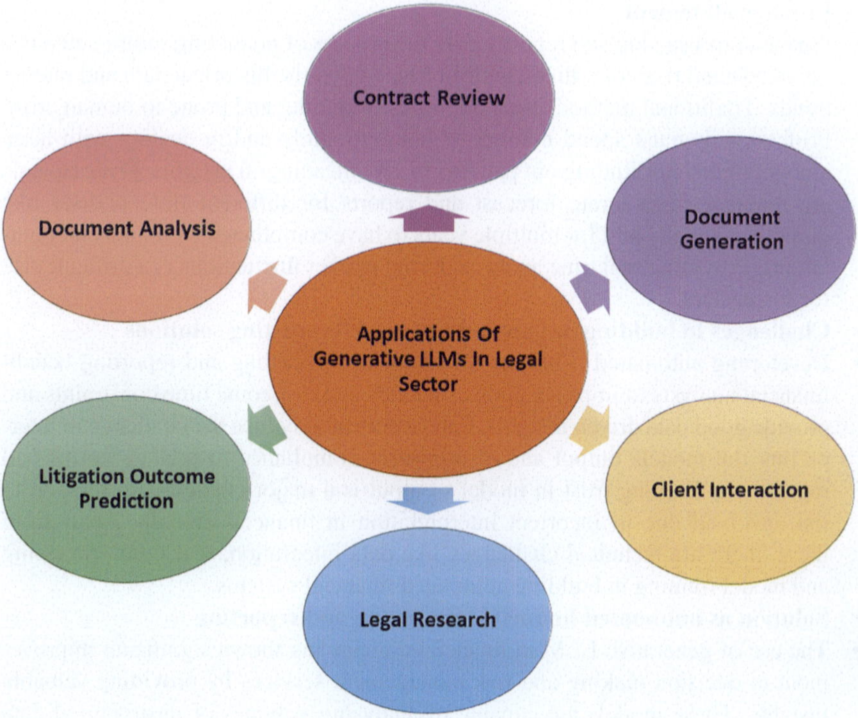

Fig. 9.6 Applications of generative LLMs in legal sector

review, document generation, legal research, litigation outcome prediction, client interaction, e-discovery and due diligence.

9.2.3 Financial Forecasting and Reporting Using Generative Models

Financial forecasting and reporting are critical aspects of any financial firm for planning and strategic decision making. Generative LLMs can analyse patterns within data to provide insights helping in forecasting and reporting. Using generative LLMs, the finance firms and professionals can optimize their workflows, enhance productivity, detect frauds, generate reports and forecast trends and risks. Firms can now take informed decisions at faster pace, in error free manner and provide high level of customer satisfaction. This section presents case study on how generative LLMs are helpful in financial forecasting and reporting.

- **Problem statement**
- Financial forecasting and reporting are the process of predicting future outcomes i.e. trends and risks of a financial firm based upon the historical data and market trends. Traditional methods used are time consuming and prone to human error. Professionals must spend number of hours to study and in dealing with large amount of data for finding out patterns to give meaningful insights. Professionals are required to generate forecast and reports for different time periods like monthly, annually and for multiple years to have compliance with financial regulations. Providing real time analysis during market fluctuations is a difficult task for forecasters.
- **Challenges in building automated financial forecasting solutions**
- Developing automated solutions for financial forecasting and reporting benefit financial analysts to improve upon efficiency and overcome time constraints and provide good data driven insights. Forecasters mainly face the challenge in interpreting the models output and also ensures compliance to privacy norms and regulations. Earning trust in model's output is a major challenge in relation to risk involved due to incorrect interpretation in financial domain. Apart from these, there are technical challenges like data integration, real time processing and model training in building automated financial systems.
- **Solution as automated financial forecasting and reporting**
- The use of generative LLM in financial domain has shown significant improvement in decision making and risk management services by providing valuable insights. These models are capable of analysing volumes of unstructured data like social media post, news articles and financial reports and forecast market trends and inform investment plans. Trained on diverse sets of data, GPT-4 has been notably used in financial analysis and can generate contextually coherent textual output. FinBERT is a BERT model specifically designed and tuned for financial domain. It is being used to capture market sentiment and financial trends to generate reports for industries. BloombergGPT is an another generative LLM trained on financial data used to create financial new articles and detailed forecasting reports.
- In this direction researchers have also applied GPT and other models in the analysis of financial documents. Number of surveys conducted by different researchers investigating how generative LLMs can improve transparency and quality of financial accounting and reporting. The studies concludes that LLMs are capable of detecting anomalies and provide valuable insights about financial disclosures [14–17]. In [18] the authors explored six different principles for the effective use of Chat GPT for accounting services provided by certified public accountants. Fotoh Elad and Tatenda Mugwira [19] explores impact of LLM on external audits and its ethical implications. They used quotes from audit firms as prompts to ChatGPT. For checking completeness and accuracy of responses, statistical methods like Cohen's kappa and Mann-Whitney U-test. In another study [20], LLMs were used to predict future earnings and helps in building investment strategies. The work concludes that LLM generated valuable insights about a company's future performance. C. Liu et al. [21] worked on sentiment analysis

in financial markets and focus on relation between market sentiments and bitcoin prices.

- **Results and benefits of using generative LLM in financial forecasting and reporting**
- Generative LLM has empowered the financial industry with the ability to generate data driven insights. This has led to significant improvement over traditional methodologies applied for financial analysis. New models are well adapted to changing market scenarios and can perform real time analysis. Reduction in total analysis time and improved prediction accuracy are the two main benefits of using generative LLMs in financial sector. Accurate forecasting help financers in taking strategic decisions at a fast pace, detect frauds and mitigate risks. Creating financial reports, bookkeeping, accounting like tasks is now automated and saves both time and effort of persons involved in these processes. The fusion of GPT-4 into financial forecasting and reporting processes has shown improvement and efficiency in the way financial data can be processed and analysed. Generative LLMs can be used in other applications apart from financial forecasting and reporting in finance sector. Few interesting applications of generative LLMs other than forecasting and reporting in financial sector are:
  - Scenario Analysis
  - Fraud Detection
  - Risk Management
  - Chatbots for Customer Support
  - Accounting Automation
  - Portfolio Optimization and Personalized Financial Advice
  - Market research

9.3 Conclusion

Amalgamation of generative LLMs and NLP has proven the potential of AI in transforming the way we can communicate with machines. Generative LLMs has refined the boundaries of human-machine interaction. Communication with machine is now more realistic, intelligent and creative. Generative LLMs provide complex and creative prompts which enhance the performance of NLP models. LLMs are capable of analysing and comprehending the meaning of text and can also generate novel contextually relevant content. The automation of content creation has paved a path of numerous possibilities of advancements in high stake domains like law, finance and healthcare industry. Businesses can exploit these technologies to improve customer interactions, product development and descriptions, marketing strategies and other content. Generative LLMs helps to streamline processes by automating routine tasks that involve language processing like writing emails, extracting insights from large volumes of data and generating reports. This not only saves time and resources but also ensures accuracy, consistency and efficiency of output. These advancements in generative technology comes with its own challenges of bias and fairness, interpretability, domain-adaptability and privacy. The generative power of LLMs should be harnessed in ethical and responsible manner to develop future NLP applications.

References

1. G. Yenduri *et al.*, "Gpt (generative pre-trained transformer)–a comprehensive review on enabling technologies, potential applications, emerging challenges, and future directions," *IEEE Access*, 2024.
2. Z. Z. Chen *et al.*, "A Survey on Large Language Models for Critical Societal Domains: Finance, Healthcare, and Law," *arXiv preprint arXiv:2405.01769*, 2024.
3. K. Subramanyam Kalyan, "A Survey of GPT-3 Family Large Language Models Including ChatGPT and GPT-4," *arXiv e-prints*, p. arXiv-2310, 2023.
4. D. Xu *et al.*, "Large language models for generative information extraction: A survey," *Front Comput Sci*, vol. 18, no. 6, p. 186357, 2024.
5. M. Alawida, S. Mejri, A. Mehmood, B. Chikhaoui, and O. Isaac Abiodun, "A comprehensive study of ChatGPT: advancements, limitations, and ethical considerations in natural language processing and cybersecurity," *Information*, vol. 14, no. 8, p. 462, 2023.
6. C. Peng *et al.*, "A study of generative large language model for medical research and healthcare," *NPJ Digit Med*, vol. 6, no. 1, p. 210, 2023.
7. Q. Lu, D. Dou, and T. Nguyen, "ClinicalT5: A generative language model for clinical text," in *Findings of the Association for Computational Linguistics: EMNLP 2022*, 2022, pp. 5436–5443.
8. R. Yang *et al.*, "Ascle—a Python natural language processing toolkit for medical text generation: development and evaluation study," *J Med Internet Res*, vol. 26, p. e60601, 2024.
9. D. Van Veen *et al.*, "Clinical text summarization: adapting large language models can outperform human experts," *Res Sq*, 2023.
10. S. Hassani, "Enhancing Legal Compliance and Regulation Analysis with Large Language Models," *arXiv preprint arXiv:2404.17522*, 2024.
11. J.-H. Huang, C.-C. Yang, Y. Shen, A. M. Pacces, and E. Kanoulas, "Optimizing numerical estimation and operational efficiency in the legal domain through large language models," in *Proceedings of the 33rd ACM International Conference on Information and Knowledge Management*, 2024, pp. 4554–4562.
12. J. Savelka and K. D. Ashley, "The unreasonable effectiveness of large language models in zero-shot semantic annotation of legal texts," *Front Artif Intell*, vol. 6, p. 1279794, 2023.
13. S. Schweitzer and M. Conrads, "The digital transformation of jurisprudence: An evaluation of ChatGPT-4's applicability to solve cases in business law," *Artif Intell Law (Dordr)*, pp. 1–25, 2024.
14. H. Zhao *et al.*, "Revolutionizing finance with llms: An overview of applications and insights," *arXiv preprint arXiv:2401.11641*, 2024.
15. D. K. C. Lee, C. Guan, Y. Yu, and Q. Ding, "A comprehensive review of generative AI in finance," *FinTech*, vol. 3, no. 3, pp. 460–478, 2024.
16. S. P. Dhake, L. Lassi, A. Hippalgaonkar, R. A. Gaidhani, and J. NM, "Impacts and Implications of Generative AI and Large Language Models: Redefining Banking Sector," *Journal of Informatics Education and Research*, vol. 4, no. 2, 2024.
17. Y. Nie *et al.*, "A Survey of Large Language Models for Financial Applications: Progress, Prospects and Challenges," *arXiv preprint arXiv:2406.11903*, 2024.
18. D. Street, J. Wilck, and Z. Chism, "Six Principles for the Effective Use of ChatGPT and other Large Language Models in Accounting," *CPA Journal (Forthcoming)*, 2023.

19. F. Elad and T. Mugwira, "Exploring Large Language Models in External Audits: Implications and Ethical Considerations," *Available at SSRN 4932003*.
20. A. Kim, M. Muhn, and V. Nikolaev, "Financial statement analysis with large language models," *arXiv preprint arXiv:2407.17866*, 2024.
21. C. Liu, A. Arulappan, R. Naha, A. Mahanti, J. Kamruzzaman, and I.-H. Ra, "Large language models and sentiment analysis in financial markets: A review, datasets and case study," *IEEE Access*, 2024.

Chapter 10
Responsible and Ethical AI in Natural Language Processing

Saurabh Raj Sangwan and Akshi Kumar

Abstract As Natural Language Processing (NLP) becomes deeply integrated into critical decision-making areas, responsible and ethical AI principles are paramount. This chapter explores the foundational principles of responsible AI—transparency, fairness, inclusivity, and accountability—and their application in NLP. It addresses key ethical challenges in fields like healthcare, finance, and law, where NLP's role can significantly impact individual and societal outcomes. The chapter examines practical techniques for bias mitigation, privacy protection, and explainability, offering solutions to improve ethical NLP development. By analyzing collaborative frameworks, sector-specific requirements, and emerging trends, this chapter highlights pathways to ensure that NLP systems serve all users equitably and adapt to evolving societal values.

As Natural Language Processing (NLP) and artificial intelligence (AI) applications expand into daily life and decision-making, their responsible and ethical implementation has become essential. NLP technologies now support tasks ranging from customer service and legal documentation to sentiment analysis and content generation. While these tools offer immense potential for positive societal impact, they also introduce ethical risks and challenges [1]. Models trained on large datasets can inadvertently perpetuate biases or reinforce stereotypes, potentially affecting real-world decisions. Furthermore, the "black box" nature of many AI systems [2] can hinder transparency, leaving users with limited understanding of how decisions are made.

S. R. Sangwan
Department of Computer Science and Engineering, Artificial Intelligence and Machine Learning, G L Bajaj Institute of Technology and Management, Greater Noida, Uttar Pradesh, India
e-mail: saurabh.sangwan@glbitm.ac.in

A. Kumar (✉)
School of Computing, Goldsmiths, University of London, London, UK
e-mail: Akshi.Kumar@gold.ac.uk

A. Kumar, S. R. Sangwan (eds.), *Transformative Natural Language Processing*,
https://doi.org/10.1007/978-3-031-88988-2_10

The principles of responsible AI seek to address these challenges by promoting fairness, transparency, inclusivity, and accountability [3]. In this chapter, we will explore how these principles apply to NLP, identifying key techniques for achieving responsible AI in practice, ethical challenges that may arise, and frameworks to guide future advancements. Through careful design and implementation, developers and researchers can build NLP systems that are not only innovative but also trustworthy and aligned with societal values.

10.1 Principles of Responsible AI in NLP Applications

Responsible AI in NLP is grounded in three foundational principles: transparency, fairness, and inclusivity [4]. These principles ensure that NLP applications operate equitably across demographic groups, offer clarity in their decision-making, and represent diverse linguistic and cultural communities. By adhering to these principles, NLP developers can minimize harm, protect user rights, and foster trust.

- **Fairness**: Fairness in NLP [5] requires that models treat all individuals and groups equitably. Since NLP applications often involve complex language data, they can unintentionally produce biased or discriminatory outcomes. Ensuring fairness requires careful handling of training data and rigorous testing to confirm that all groups are represented adequately. For instance, if a chatbot designed for customer service consistently misinterprets requests from speakers with regional dialects, it may be perpetuating unfair treatment.
- **Transparency**: Transparency entails making the workings of NLP models understandable to both users and stakeholders [6]. This can involve explaining how predictions are made, what influences a model's decisions, and disclosing any known limitations. For example, when an AI system suggests legal advice, transparency in its reasoning allows legal professionals to validate its recommendations, ensuring they are appropriate for the context.
- **Inclusivity and Accessibility**: Inclusivity ensures that NLP models represent diverse linguistic [7], cultural, and demographic perspectives, allowing all users to benefit. Accessibility means designing NLP systems that can be easily used by people with different abilities and backgrounds. For example, a sentiment analysis model that works accurately for all major dialects within a language, including African American Vernacular English (AAVE) [8], increases inclusivity and prevents the exclusion of certain communities.

10.1.1 Ensuring Transparency and Fairness in AI Models

Ensuring transparency and fairness in NLP models is fundamental for creating systems that users can understand and trust. Transparency allows users to see how inputs affect outputs, while fairness prevents biased outcomes that can harm individuals or groups. The following techniques support transparency and fairness in NLP applications:

- **Transparency Techniques**

  (a) *Explainable AI (XAI):* Explainable AI [9] provides tools for understanding and visualizing how a model arrives at its outputs. Methods like Layer-wise Relevance Propagation (LRP) [10] and Integrated Gradients [11] highlight important words or phrases in input data that contribute to the model's prediction. In an NLP application for medical diagnosis, for example, these tools can help healthcare providers understand which symptoms or keywords led the model to flag a particular condition, allowing them to assess the relevance of the prediction.

  (b) *Attention Mechanisms:* Common in transformer-based models like BERT [12] and GPT [13], attention mechanisms can reveal which parts of an input the model is focusing on. By visualizing attention weights, developers can observe which words or phrases influenced the prediction. In sentiment analysis, attention mechanisms can show how certain words affect the model's interpretation of sentiment, providing users with insight into the model's decision-making process.

  (c) *Interpretable Embeddings:* Word embeddings, which represent words as dense vectors, can be difficult to interpret. Techniques such as Principal Component Analysis (PCA) and t-SNE (t-distributed Stochastic Neighbour Embedding) allow developers to visualize and interpret relationships between words [14]. For example, if a model embeds the word "doctor" closer to "male" than "female," it may indicate an underlying bias. Visualizing embeddings can help developers detect and address these biases.

  (d) *Model Cards and Datasheets for Datasets:* Model cards document a model's design, intended use cases, limitations, and known biases. Datasheets describe a dataset's characteristics, origin, collection methods, and potential biases [15]. By making this information accessible, developers and users can make informed decisions about how to use the model responsibly. For example, a model card might indicate that an NLP model trained primarily on Western media sources may not generalize well to non-Western contexts.

- **Fairness Techniques**

  (a) *Bias Detection and Mitigation Tools:* Tools like IBM's AI Fairness 360 and Google's What-If Tool help identify and quantify bias in models and datasets [16]. These tools can analyze model predictions across demographic groups, highlighting disparities. For example, if an NLP model shows a tendency to interpret positive language more often when used by certain demographic groups, this disparity can be corrected through retraining or rebalancing techniques.

  (b) *Adversarial Debiasing:* This technique trains a secondary model, called an adversary, to detect biased patterns in the main model's predictions [17]. The primary model is then optimized to reduce these patterns, resulting in fairer outputs. In hiring applications, adversarial debiasing can help ensure that gendered language does not influence job candidate evaluations, promoting more equitable hiring practices.

(c) *Counterfactual Data Augmentation:* Counterfactual data augmentation [18] creates new examples by changing certain attributes to test for fairness. For example, by replacing male pronouns with female pronouns in sentences and checking model predictions, developers can detect and reduce gender bias.

(d) *Fairness Constraints and Regularization:* Fairness constraints and regularization terms added during training can reduce the influence of sensitive attributes on predictions [19]. For instance, in a financial NLP model, regularization can prevent the system from over-relying on features like zip codes, which may correlate with socio-economic status, helping to reduce unfair outcomes.

10.1.2 Mitigating Bias and Promoting Inclusivity

To mitigate bias and promote inclusivity, NLP applications should be designed to serve diverse user groups equitably. Bias often arises from imbalanced datasets or from underrepresentation of certain groups, which can lead to predictions that are unfair or inaccurate [20]. Mitigating bias and promoting inclusivity requires thoughtful data collection, rigorous testing, and specific bias-reduction techniques.

- **Bias Mitigation Techniques**

(a) *Data Rebalancing and Reweighting:* Data rebalancing techniques, such as oversampling or reweighting underrepresented groups, provide fairer model predictions. In sentiment analysis, for example, rebalancing a dataset that is skewed toward positive reviews can improve the model's ability to recognize negative sentiments accurately. This technique also helps address biases that may arise from underrepresented populations in training data.

(b) *Bias-aware Preprocessing:* Preprocessing methods, such as de-biasing word embeddings, can help mitigate biased associations. For example, techniques like Hard Debiasing shift word vectors so that terms like "doctor" or "engineer" are not associated with a particular gender. This method is effective in reducing stereotypical associations in word embeddings, improving the fairness of downstream NLP models.

(c) *Adversarial Training for Bias Detection:* Adversarial training involves using a secondary model to detect biased predictions, allowing the primary model to be refined. This approach is effective in hate speech detection, where models need to distinguish between offensive language and legitimate expressions used by marginalized groups. By training the primary model to avoid bias in specific cases flagged by the adversary, developers can build more accurate and fair models.

(d) *Equalized Odds Post-processing:* Equalized odds post-processing adjusts prediction probabilities for different groups to meet fairness criteria, ensuring a fairer distribution of outcomes. This method is particularly useful in classification tasks where disparities between groups may occur. For example, in a credit scoring model, this technique helps to avoid any demographic group being unfairly favored or penalized by the model's predictions.

- **Inclusivity Techniques**

(a) *Language and Dialect Inclusion:* NLP models should be designed to accommodate multiple languages and dialects, supporting global users. Multilingual training and fine-tuning on diverse datasets allow NLP systems to better understand non-standard dialects. For instance, a sentiment analysis model fine-tuned on African American Vernacular English (AAVE) and other dialects alongside standard English can reduce misinterpretations and improve inclusivity.

(b) *Inclusive Dataset Collection:* Inclusivity begins with collecting diverse datasets that represent various linguistic, cultural, and socio-economic backgrounds. Projects like Mozilla's Common Voice invite users to contribute samples in different languages and dialects, building a dataset that reflects the diversity of real-world language use and helping voice recognition technologies become more accessible globally.

(c) *User-Centric Design and Feedback:* Designing NLP systems with user feedback from diverse groups ensures that applications are relevant and fair. For example, mental health chatbots can be adapted to recognize culturally sensitive language, allowing them to better support users from different backgrounds.

(d) *Regular Audits for Inclusivity:* Regularly auditing NLP models is essential to maintaining inclusivity as languages and societal norms evolve. Periodic evaluations can reveal whether specific groups are consistently misrepresented or misunderstood, prompting updates or retraining. External audits or bias review panels can also be valuable for ensuring inclusivity in real-world applications.

Example of Inclusivity in Practice: Google Translate's addition of new languages and dialects highlights the importance of inclusivity in NLP. By expanding support for underrepresented languages, Google has increased access to its translation tools for global users, allowing more people to benefit from NLP technology.

Thus, ensuring transparency, fairness, bias mitigation, and inclusivity are critical in developing responsible NLP applications. By adopting these techniques, developers can create models that align with ethical standards, promoting a fair and inclusive experience for users across different backgrounds. Through responsible AI principles, NLP systems can uphold trust, foster equity, and remain adaptable to future societal changes, ensuring that AI technology serves all of humanity.

10.2 Ethical Challenges and Solutions in NLP

As NLP applications expand across critical sectors such as healthcare, law, and finance, ethical challenges specific to each field demand attention. High-stakes NLP applications have the power to influence medical diagnoses, legal decisions, and financial recommendations, which makes it essential to address potential ethical pitfalls. These include issues of privacy, accountability, fairness, and the risk of harm through inaccurate or biased outputs. Ethical solutions are necessary to ensure that NLP applications do not inadvertently cause harm or infringe upon rights. This section outlines key ethical concerns and solutions in medical and legal NLP applications and explores governance and oversight practices in financial NLP.

10.2.1 Addressing Ethical Concerns in Medical and Legal NLP

In the medical and legal sectors, NLP applications must prioritize ethics due to the sensitive nature of data and the potential impact on individuals' lives. Ethical challenges in these fields often relate to privacy, accuracy, accountability, and trust. For example, an NLP model used in medical diagnostics might analyze patient records to predict disease risk, making the privacy and accuracy of its outputs crucial. Similarly, in legal NLP, a tool designed to assist with case research or contract analysis must avoid misinterpretation of language that could result in flawed legal judgments. Addressing these ethical concerns requires a blend of technical and regulatory solutions.

- **Privacy and Data Protection:** Medical and legal data is often highly sensitive, containing personal information that is protected under regulations like the Health Insurance Portability and Accountability Act (HIPAA) in the United States and the General Data Protection Regulation (GDPR) in the European Union. NLP applications in these fields must adhere to strict privacy requirements to prevent unauthorized access or misuse of data. Techniques such as data anonymization and pseudonymization are widely used to safeguard patient and client information. In data anonymization, identifiable information is removed or altered, making it difficult to trace data back to specific individuals. For example, a medical NLP tool analyzing patient histories would anonymize names, addresses, and specific identifiers, ensuring that data privacy is preserved. Another approach to privacy in medical NLP is federated learning, where models are trained across multiple locations using local data without transferring sensitive information to a central server. For instance, federated learning could be employed across hospitals, enabling each hospital to contribute to an NLP model without sharing raw patient data. This decentralized training approach helps protect patient privacy while allowing institutions to benefit from collective data insights.

- **Accuracy and Accountability:** Accuracy in medical and legal NLP is vital, as errors can lead to harmful consequences. For example, a medical NLP system used for disease prediction must deliver accurate results, as an incorrect prediction could result in unnecessary treatments or overlooked diagnoses. In legal applications, an inaccurate NLP-based contract analyzer could misinterpret clauses, leading to flawed contract terms or incorrect legal advice. To mitigate these risks, rigorous testing and validation on high-quality datasets is essential before deploying NLP models in these domains. In medical NLP, for example, accuracy can be improved by training models on large, clinically verified datasets and by continually monitoring performance metrics such as sensitivity and specificity to ensure reliable outputs. Accountability mechanisms are also critical. By implementing audit trails and version control, developers can track changes to models, creating a record of modifications that may affect outcomes. For instance, if a medical NLP model is updated, an audit trail can help ensure that the new version maintains high accuracy and that any discrepancies are noted and evaluated. Similarly, in legal applications, an audit trail can help track the reasoning behind model decisions, such as why it classified a particular case as relevant or suggested a particular legal phrase, providing transparency and accountability.
- **Bias Mitigation in Sensitive Domains:** In both medical and legal fields, NLP applications must avoid biases that could lead to discriminatory or unfair outcomes. For instance, if a medical NLP system is trained predominantly on data from a particular demographic group, it may not perform well for patients outside that group, leading to healthcare disparities. Similarly, legal NLP tools must be cautious of biases embedded in historical case data that may reflect outdated or unjust societal norms. Bias mitigation techniques, such as reweighting underrepresented groups in training data or debiasing embeddings, are crucial to creating equitable NLP applications. In practice, algorithmic fairness frameworks like Equalized Odds and Demographic Parity can be applied in medical and legal NLP to ensure fair treatment across demographic groups. Equalized Odds ensures that model outputs are equally likely across protected groups, such as different races or genders, preventing one group from being disadvantaged. For example, in a medical risk prediction model, Equalized Odds would require that the model's sensitivity and specificity are consistent across racial groups, reducing health disparities.

As an example, consider an NLP model used in medical transcription for clinical notes. This model must accurately recognize medical terminology across different dialects and accents to avoid potential inaccuracies. To prevent bias, developers could include diverse speech samples from multiple demographics in the training data, ensuring the model performs equally well for all patients. By implementing such inclusive data practices and bias checks, developers help prevent inequitable treatment of patients based on language or accent, promoting fairer healthcare outcomes.

10.2.2 Governance and Ethical Oversight in Financial NLP Applications

Financial NLP applications introduce unique ethical concerns, particularly around fairness, privacy, accountability, and explainability. As these tools are increasingly used for risk assessment, sentiment analysis, and automated customer service, robust governance and ethical oversight become essential to prevent unethical practices, ensure data protection, and maintain market fairness.

- **Privacy and Data Security:** Financial data is often personal and sensitive, requiring adherence to privacy regulations like GDPR and Payment Card Industry Data Security Standard (PCI DSS). Privacy techniques such as encryption and data masking are crucial for protecting financial data in NLP applications. Encryption secures data by converting it into a code, making it accessible only to authorized users. In data masking, sensitive information is replaced or obfuscated so that unauthorized users cannot view personal data. Differential Privacy is another technique increasingly used in financial NLP to maintain privacy while analyzing large datasets. For instance, a financial NLP model might analyze spending trends across thousands of transactions without exposing any individual's data. Differential privacy achieves this by adding a small amount of noise to the data, allowing analysis of aggregate trends without revealing specifics. This approach is particularly useful in scenarios where data is shared with third-party analytics firms or across departments.
- **Algorithmic Transparency and Explainability:** Financial NLP models must prioritize transparency, as decisions regarding credit scores, loan approvals, and investment risks can significantly impact individuals. Explainable AI (XAI) techniques are essential in these applications to provide clarity on why certain decisions were made. For example, an NLP-based credit scoring model that considers social media activity might highlight key phrases or sentiment scores that influenced the credit decision. Techniques like SHAP (SHapley Additive exPlanations) and LIME (Local Interpretable Model-agnostic Explanations) provide insights into model outputs, helping users understand the factors driving each decision. Explainability is also vital in maintaining public trust and regulatory compliance. For example, regulatory bodies like the European Banking Authority (EBA) mandate transparency in automated credit scoring processes to ensure fair treatment of applicants. Providing explainable insights into NLP models allows financial institutions to meet such regulatory standards and helps users understand why certain financial decisions are made.
- **Fairness and Bias Prevention:** Bias in financial NLP models can lead to unfair treatment of specific demographic groups, which is particularly concerning in credit scoring, fraud detection, and investment recommendations. Bias mitigation strategies are essential to ensure fair and inclusive treatment. For instance, a credit scoring model trained predominantly on data from high-income individuals may perform poorly for lower-income applicants. Techniques such as re-

sampling and reweighting underrepresented data, as well as fairness constraints like Demographic Parity, are effective in correcting such imbalances. Demographic Parity, for example, ensures that all groups have equal likelihoods of receiving a positive outcome, such as loan approval, regardless of their background. Additionally, Fair Lending Practices frameworks guide financial institutions in minimizing bias in credit and loan decisions. The Equal Credit Opportunity Act (ECOA) in the U.S. mandates that lenders treat all applicants fairly, prohibiting discrimination based on race, gender, religion, and other protected characteristics. NLP developers working in financial domains can use adversarial debiasing to further ensure that their models comply with these regulations. In adversarial debiasing, an adversarial model actively monitors predictions, flagging any indications of bias that the main model can then correct, promoting fairness in credit assessments.

- **Accountability and Regulatory Compliance:** Accountability mechanisms are essential in financial NLP to ensure ethical practices. **Audit trails** and **version control** allow financial institutions to track changes in models and ensure compliance with regulatory standards. For example, an audit trail in a fraud detection system can record model decisions, inputs, and modifications over time, helping investigators review how a suspicious transaction was flagged. This level of traceability not only aids in compliance with financial regulations but also facilitates post-hoc analysis in case of errors or disputes.

For example, a financial institution using NLP to screen loan applications could implement a comprehensive ethical framework to address bias, transparency, and accountability. Suppose this institution uses sentiment analysis on applicants' online reviews to gauge risk. To ensure fairness, the institution might use adversarial debiasing to prevent biases from affecting the sentiment scores, ensuring equal treatment across demographics. The institution could also apply explainability techniques like SHAP to show why certain phrases influenced the sentiment score. Finally, an audit trail can record each decision, providing a complete record that supports accountability and compliance with financial regulations.

- **Ethical Oversight Boards and Committees:** Financial institutions are increasingly establishing ethical oversight boards or AI ethics committees to monitor the ethical implementation of NLP applications. These boards include experts in data ethics, finance, and technology, who review and approve NLP models before they are deployed. Ethical oversight boards ensure that models comply with both regulatory standards and internal ethical policies, minimizing risks of harm and maintaining ethical practices in the use of financial data.

Addressing ethical concerns in high-stakes domains like healthcare, law, and finance requires a combination of technical solutions, regulatory compliance, and organizational accountability. By implementing privacy-preserving techniques, explainable AI, bias mitigation strategies, and robust accountability frameworks, NLP applications can uphold ethical standards and reduce potential harm. Governance and ethical oversight practices, such as establishing ethics boards and adhering to regulatory requirements, further ensure that NLP applications align with societal expectations and contribute positively to their respective domains. The Table 10.1 below summarizes the primary ethical challenges in medical, legal, and financial NLP applications, along with potential impacts, solutions, and relevant frameworks. By adopting these practices, developers and organizations can address the unique ethical concerns that arise in each sector, fostering responsible and trustworthy NLP systems.

10.3 Pathways to Ethical AI Development

As the applications of NLP expand and become more integrated into society, the need for ethically grounded, responsible AI systems is increasingly pressing. The complexities of language, the potential for AI to influence behaviour, and the risks of bias necessitate frameworks and future-oriented approaches that support ethical development. This section explores collaborative frameworks that bring together diverse expertise for responsible NLP research and discusses future directions in responsible AI to create NLP applications that are trustworthy, inclusive, and adaptable to societal shifts.

10.3.1 Collaborative Frameworks for Ethical NLP Research

Ethical AI development in NLP requires interdisciplinary collaboration that brings together researchers, industry professionals, ethicists, and policymakers [21]. Such frameworks promote the creation of systems that prioritize societal welfare, accountability, and inclusivity. Collaborative frameworks provide avenues for shared learning, help standardize ethical practices and ensure that all perspectives are considered when addressing the ethical complexities of NLP.

- **Interdisciplinary Research Teams:** Collaboration across disciplines ensures that ethical NLP models are developed with a comprehensive understanding of social, cultural, and psychological factors that influence language. For instance, a collaboration between NLP researchers and social scientists can help identify biases embedded in linguistic data, allowing teams to address these biases proactively. In developing a mental health chatbot, a team comprising NLP specialists, psychologists, and sociolinguists can design a system that responds empathetically and avoids harmful language, ensuring safe interaction for users.
- **Public-Private Partnerships:** Partnerships between academia, government agencies, and the private sector drive responsible innovation in NLP by pooling resources and expertise to address common ethical challenges. Organizations like Partnership on AI[1] facilitate knowledge-sharing by bringing together leading technology firms, research institutions, and nonprofits to develop and promote ethical AI practices. For example, through such partnerships, financial NLP tools can be evaluated for fairness in loan decision-making, with government bodies providing regulatory guidance and tech firms contributing data expertise.

[1] https://partnershiponai.org/

Table 10.1 Key ethical challenges and solutions in high-stakes NLP applications

Sector	Ethical challenge	Potential impact	Solution/technique	Example/ framework
Medical NLP	Privacy and Data Protection	Risk of data breaches and privacy violations	Data Anonymization, Federated Learning	Compliance with HIPAA (US), GDPR (EU)
	Accuracy and Reliability	Incorrect diagnoses or treatment recommendations	Rigorous Testing, High-Quality Dataset Curation	Sensitivity and Specificity Testing
	Bias and Fairness	Health disparities among demographic groups	Reweighting, Equalized Odds, Diverse Training Data	Equalized Odds for demographic parity
	Accountability	Difficulty tracing model errors in clinical applications	Audit Trails, Version Control	Medical AI Transparency Act (proposed policies)
Legal NLP	Privacy and Confidentiality	Risk of exposing confidential client information	Data Masking, Controlled Access Protocols	GDPR (EU), Attorney-Client Privilege
	Misinterpretation of Language	Incorrect legal advice or flawed contract terms	Explainable AI, Multi-layer Validation	Layer-wise Relevance Propagation (LRP)
	Bias from Historical Data	Potential reinforcement of outdated or biased legal norms	Algorithmic Fairness, Debiasing Techniques	Counterfactual Data Augmentation
	Accountability	Lack of transparency in AI-generated legal advice	Audit Trails, Model Documentation (Model Cards)	Model Cards for AI in LegalTech
Financial NLP	Privacy and Security of Financial Data	Risk of unauthorized access and financial fraud	Encryption, Differential Privacy, Data Masking	PCI DSS, GDPR (EU)
	Fairness in Credit and Loan Decision-Making	Discrimination against certain demographic groups	Re-sampling, Demographic Parity, Adversarial Debiasing	Equal Credit Opportunity Act (ECOA) (US)
	Algorithmic Transparency	Lack of understanding in automated credit or risk assessment	Explainable AI (SHAP, LIME)	Explainable AI mandates by European Banking Authority (EBA)
	Accountability and Compliance	Regulatory non-compliance, lack of accountability in decision-making	Ethical Oversight Boards, Audit Trails	Establishment of AI Ethics Committees

- **Standardized Ethics Guidelines and Toolkits:** Shared ethical guidelines and toolkits ensure that ethical considerations are consistent across NLP projects, even when stakeholders come from different backgrounds. The Ethics Guidelines for Trustworthy AI by the European Commission[2] provides a standardized framework for ethical AI development, outlining principles like fairness, accountability, and transparency. Similarly, the AI Fairness 360 Toolkit by IBM[3] helps organizations detect and mitigate bias, offering open-source tools that ensure adherence to fairness standards in NLP applications.
- **Data and Model Documentation:** Proper documentation practices like Model Cards [22] and Datasheets for Datasets [23] are critical to promoting ethical NLP research. Model cards, which document a model's capabilities, limitations, and ethical considerations, provide transparency and inform users about how the model was developed and tested. For instance, a model card for a legal text summarization tool would include information on the datasets used, any known biases, and limitations in handling specialized legal terminology. This transparency is especially important in NLP applications where users may rely heavily on the tool's outputs for professional decision-making.
- **Ethics Review Committees:** Ethical oversight committees within organizations and research institutions provide a structured approach to evaluating NLP projects for ethical compliance. These committees typically include ethicists, legal advisors, and subject matter experts who review projects before deployment, ensuring that ethical standards are met. For example, an ethics review committee for a medical NLP project might evaluate whether the model adheres to patient privacy laws and verify that the data is anonymized appropriately. Figure 10.1 visually maps out these stakeholders and their interactions, providing a conceptual overview that underpins the collaborative approaches discussed in this section.

10.3.2 Example of Collaborative Frameworks in Action: The Common Voice Project

The Mozilla Common Voice Project exemplifies how collaboration can drive inclusivity in NLP. This initiative collects voice data from volunteers worldwide, representing a diverse range of languages and dialects, which helps develop more inclusive voice recognition models. By collaborating with communities, Mozilla ensures that underrepresented languages, including indigenous and regional dialects, are included in NLP systems, promoting linguistic diversity and equity in AI (Table 10.2).

[2] https://digital-strategy.ec.europa.eu/en/library/ethics-guidelines-trustworthy-ai

[3] https://aif360.res.ibm.com/

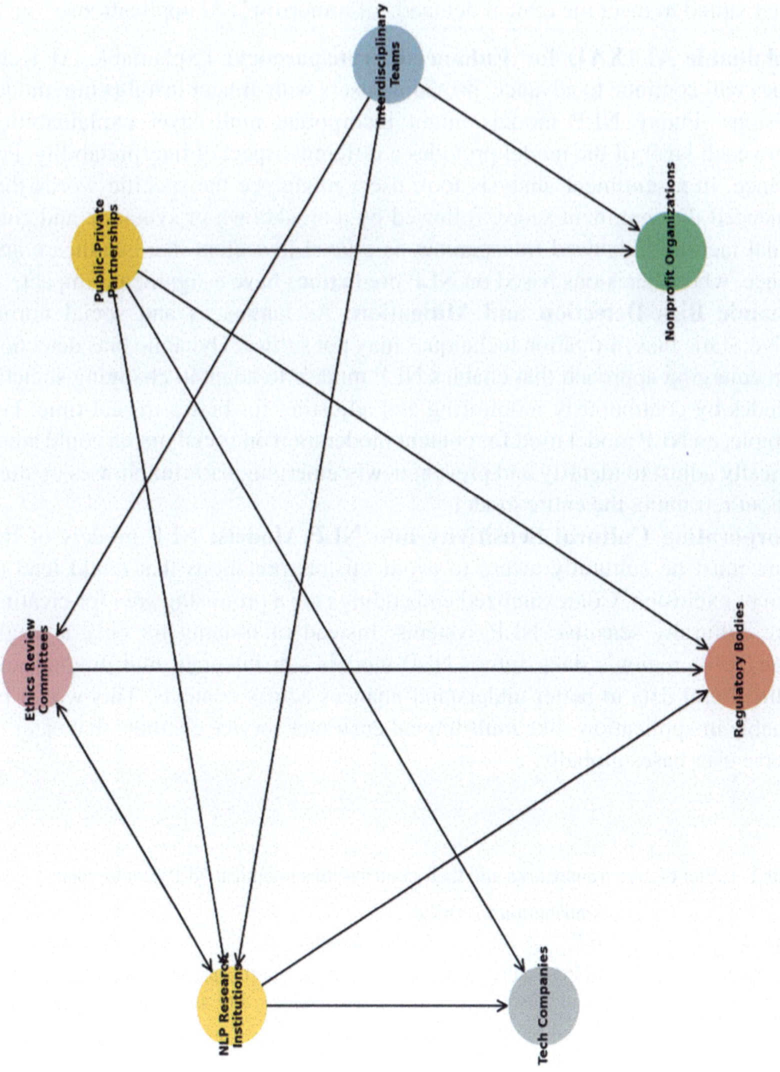

Fig. 10.1 An illustration of collaborative frameworks in ethical NLP research

10.3.3 Future Directions in Responsible AI

The future of responsible AI in NLP involves the development of advanced techniques for transparency, adaptability to social changes, and proactive bias prevention. By embracing innovative approaches, NLP developers can build systems that are better suited to meet the ethical demands of tomorrow's AI applications.

- **Explainable AI (XAI) for Enhanced Transparency:** Explainable AI techniques will continue to advance, providing users with greater insights into model decisions. Future NLP models might incorporate multi-layer explainability, where each layer of the model provides a different aspect of interpretability. For instance, in a sentiment analysis tool, users might see the specific words that influenced the sentiment score, followed by a breakdown of syntactic and contextual factors. Enhanced transparency is crucial in sectors like healthcare and finance, where decisions based on NLP predictions have a significant impact.
- **Dynamic Bias Detection and Mitigation:** As languages and social norms evolve, static bias mitigation techniques may not suffice. Dynamic bias detection is an emerging approach that enables NLP models to adapt to changing societal attitudes by continuously monitoring and adjusting for biases in real-time. For example, an NLP model used for content moderation on social media could automatically adjust to identify and prevent newly emerging harmful phrases or slurs without retraining the entire model.
- **Incorporating Cultural Sensitivity into NLP Models:** NLP models of the future must be culturally aware to avoid misinterpretations that could lead to harm or exclusion. Contextualized embeddings are a promising area for creating more culturally sensitive NLP systems. Instead of training on only a single language or region's data, future NLP models can integrate multilingual and multicultural data to better understand nuances across contexts. This would be valuable in applications like multilingual customer service chatbots that cater to diverse user bases globally.

Table 10.2 Collaborative frameworks and their contributions to ethical NLP development

Framework	Contribution to ethical NLP	Example
Interdisciplinary Research Teams	Address cultural and social biases	Mental health chatbot with NLP, psychology, and sociolinguistics expertise
Public-Private Partnerships	Shared resources and regulatory guidance	Partnership on AI—promoting fairness in financial NLP
Standardized Ethics Guidelines	Consistent ethical practices across projects	European Commission's Trustworthy AI Guidelines
Data and Model Documentation	Transparency about model limitations	Model cards in legal NLP summarization tools
Ethics Review Committees	Structured ethical oversight	Review committee for medical NLP projects

- **Ethical AI Metrics and Benchmarks:** The development of standardized metrics for ethical performance, such as Fairness F1 Score or Bias Impact Scores, will allow researchers to assess and compare models based on ethical performance. Such metrics would go beyond traditional accuracy measures, evaluating models on criteria such as inclusivity, fairness, and privacy protection. For instance, a Bias Impact Score could quantify the extent to which a model's predictions differ across demographic groups, providing a concrete measure for evaluating fairness in NLP models.
- **Regulatory and Policy Alignment:** With increasing scrutiny of AI's societal impact, governments worldwide are developing AI policies to ensure ethical practices. NLP developers can expect future regulations to include stricter guidelines on data privacy, transparency, and accountability. For example, the European Union's proposed AI Act aims to classify AI applications based on their risk levels, with higher-risk applications requiring extensive documentation, transparency, and fairness checks. Adapting NLP research to meet regulatory standards early will be crucial for compliance and responsible deployment.

10.3.4 Example of Future-Oriented Ethical Practice: Bias Auditing in Social Media NLP

An innovative direction in responsible AI involves regular bias auditing for NLP models used in social media content moderation. These audits can periodically assess a model's performance across different demographic groups to ensure fair treatment. For example, a content moderation NLP model that flags offensive content could be audited to verify that it is equally sensitive to harmful language across various cultural contexts. This practice can prevent biases that might lead to the disproportionate removal of content from certain groups, promoting fairness and trust in AI-powered platforms (Table 10.3).

Table 10.3 Future directions in ethical AI development for NLP

Future direction	Description	Example application
Explainable AI (XAI) for Transparency	Multi-layer explanations for complex models	Sentiment analysis in finance with syntactic and contextual breakdowns
Dynamic Bias Detection and Mitigation	Real-time adjustments for emerging biases and harmful phrases	Social media content moderation
Culturally Sensitive NLP Models	Contextualized embeddings that understand cultural and linguistic nuances	Multilingual customer service chatbots
Ethical AI Metrics and Benchmarks	New metrics to assess fairness, inclusivity, and privacy alongside traditional accuracy	Bias Impact Score for demographic parity
Regulatory and Policy Alignment	Developing models that comply with emerging AI policies and regulations	European Union AI Act for high-risk applications

10.4 Conclusion

The rapid integration of NLP into high-stakes sectors such as healthcare, finance, and law has raised essential ethical concerns, underscoring the need for responsible and accountable AI systems. This chapter has explored these challenges and proposed solutions to ensure fairness, transparency, and inclusivity in NLP applications. By adopting collaborative frameworks and embracing emerging trends in explainable and culturally sensitive AI, developers can create systems that uphold ethical standards and societal values. Moving forward, ongoing advancements in transparency, regulatory alignment, and real-time bias mitigation will be essential to address the evolving ethical landscape in NLP. This commitment to ethical AI development ensures that NLP technologies foster trust, support equity, and remain adaptable to diverse cultural and social contexts, thereby promoting a fairer and more inclusive future for all users.

References

1. Tokayev, K. J. (2023). Ethical implications of large language models a multidimensional exploration of societal, economic, and technical concerns. *International Journal of Social Analytics*, *8*(9), 17-33.
2. Wischmeyer, T. (2020). Artificial intelligence and transparency: opening the black box. *Regulating artificial intelligence*, 75-101.
3. Akinrinola, O., Okoye, C. C., Ofodile, O. C., & Ugochukwu, C. E. (2024). Navigating and reviewing ethical dilemmas in AI development: Strategies for transparency, fairness, and accountability. *GSC Advanced Research and Reviews*, *18*(3), 050-058.
4. Behera, R. K., Bala, P. K., Rana, N. P., & Irani, Z. (2023). Responsible natural language processing: A principlist framework for social benefits. *Technological Forecasting and Social Change*, *188*, 122306.
5. Jourdan, F. (2024). Advancing Fairness in Natural Language Processing: From Traditional Methods to Explainability. *arXiv preprint arXiv:2410.12511*.
6. Felzmann, H., Fosch-Villaronga, E., Lutz, C., & Tamò-Larrieux, A. (2020). Towards transparency by design for artificial intelligence. *Science and engineering ethics*, *26*(6), 3333-3361.
7. Nee, J., Macfarlane Smith, G., Sheares, A., & Rustagi, I. (2021, October). Advancing social justice through linguistic justice: Strategies for building equity fluent NLP technology. In *Proceedings of the 1st ACM Conference on Equity and Access in Algorithms, Mechanisms, and Optimization* (pp. 1-9).
8. Champion, T. B., Cobb-Roberts, D., & Bland-Stewart, L. (2012). Future educators' perceptions of African American Vernacular English (AAVE). *Online Journal of Education Research*, *1*(5), 80-89.
9. Dwivedi, R., Dave, D., Naik, H., Singhal, S., Omer, R., Patel, P., ... & Ranjan, R. (2023). Explainable AI (XAI): Core ideas, techniques, and solutions. *ACM Computing Surveys*, *55*(9), 1-33.
10. Montavon, G., Binder, A., Lapuschkin, S., Samek, W., & Müller, K. R. (2019). Layer-wise relevance propagation: an overview. *Explainable AI: interpreting, explaining and visualizing deep learning*, 193-209.
11. Wang, Y., Zhang, T., Guo, X., & Shen, Z. (2024). Gradient based Feature Attribution in Explainable AI: A Technical Review. *arXiv preprint arXiv:2403.10415*.

12. Devlin, J., Chang, M. W., Lee, K., & Toutanova, K. (2018). Bert: Bidirectional encoder representations from transformers. *arXiv preprint arXiv:1810.04805*, 15.
13. Achiam, J., Adler, S., Agarwal, S., Ahmad, L., Akkaya, I., Aleman, F. L., ... & McGrew, B. (2023). Gpt-4 technical report. *arXiv preprint arXiv:2303.08774*.
14. Shahroudnejad, A. (2021). A survey on understanding, visualizations, and explanation of deep neural networks. *arXiv preprint arXiv:2102.01792*.
15. Roman, A. C., Vaughan, J. W., See, V., Ballard, S., Torres, J., Robinson, C., & Ferres, J. M. L. (2023). Open datasheets: machine-readable documentation for open datasets and responsible AI assessments. *arXiv preprint arXiv:2312.06153*.
16. Thompson, J. (2021). Mental models and interpretability in AI fairness tools and code environments. In *HCI International 2021-Late Breaking Papers: Multimodality, eXtended Reality, and Artificial Intelligence: 23rd HCI International Conference, HCII 2021, Virtual Event, July 24–29, 2021, Proceedings 23* (pp. 574-585). Springer International Publishing.
17. Aïvodji, U., Bidet, F., Gambs, S., Ngueveu, R. C., & Tapp, A. (2021). Local data debiasing for fairness based on generative adversarial training. *Algorithms*, *14*(3), 87.
18. Kacprzak, I., Glocker, B., Monteiro, M., Ribeiro, F. D. S., Xia, T., & Kainz, B. (2023). Counterfactual Data Augmentation for Deep Learning Predictive Models.
19. Zafar, M. B., Valera, I., Rogriguez, M. G., & Gummadi, K. P. (2017, April). Fairness constraints: Mechanisms for fair classification. In *Artificial intelligence and statistics* (pp. 962-970). PMLR.
20. Stranisci, M. A. (2024). Semantic-Aware Methods for the Analysis of Bias and Underrepresentation in Language Resources.
21. Lancon, E. (2022). Harmonizing AI Advancements: Fostering Interdisciplinary Synergy in Technology.
22. Mitchell, M., Wu, S., Zaldivar, A., Barnes, P., Vasserman, L., Hutchinson, B., ... & Gebru, T. (2019, January). Model cards for model reporting. In *Proceedings of the conference on fairness, accountability, and transparency* (pp. 220-229).
23. Gebru, T., Morgenstern, J., Vecchione, B., Vaughan, J. W., Wallach, H., Iii, H. D., & Crawford, K. (2021). Datasheets for datasets. *Communications of the ACM*, *64*(12), 86-92.

The manufacturer's authorised representative in the EU is Springer Nature Customer Service Centre GmbH, Europaplatz 3, 69115 Heidelberg, Germany. If you have any concerns regarding our products, please contact ProductSafety@springernature.com

Printed and bound by CPI Group (UK) Ltd, Croydon, CR0 4YY
01/10/2026
02225960-0009